INDUSTRIAL MATERIALS

VOLUME 1

Metals and Alloys

David A. Colling
University of Massachusetts Lowell

Thomas Vasilos
University of Massachusetts Lowell

Prentice Hall
Englewood Cliffs, New Jersey   Columbus, Ohio

Library of Congress Cataloging-in-Publication Data

Colling, David A.
Industrial materials.

Includes index.
Contents: v. 1. Metals and alloys—
    v. 2. Polymers, ceramics, and composites.
    1. Materials.   2. Materials—Case studies.
    3. Manufacturing processes—Case studies.
I. Vasilos, Thomas. II. Title.
TA403.C585   1995   620.1'1   94-10077
ISBN 0-02-323560-8 (v. 1)

Cover photo: David A. Colling
Editor: Stephen Helba
Production Editor: Mary Ann Hopper
Text Designer: Julia Zonneveld Van Hook
Cover Designer: Julia Zonneveld Van Hook
Production Buyer: Patricia A. Tonneman
Electronic Text Management: Marilyn Wilson Phelps, Matthew Williams, Jane Lopez,
  Karen L. Bretz
Illustrations: Diphrent Strokes, Inc.

This book was set in Times by Prentice Hall and was printed and bound by R. R. Donnelley &
Sons Company. The cover was printed by Phoenix Color Corp.

 © 1995 by Prentice-Hall, Inc.
A Simon & Schuster Company
Englewood Cliffs, New Jersey 07632

Printed in the United States of America

10 9 8 7 6 5 4 3 2 1

ISBN: 0-02-323560-8

Prentice-Hall International (UK) Limited, *London*
Prentice-Hall of Australia Pty. Limited, *Sydney*
Prentice-Hall Canada Inc., *Toronto*
Prentice-Hall Hispanoamericana, S. A., *Mexico*
Prentice-Hall of India Private Limited, *New Delhi*
Prentice-Hall of Japan, Inc., *Tokyo*
Simon & Schuster Asia Pte. Ltd., *Singapore*
Editora Prentice-Hall do Brasil, Ltda., *Rio de Janeiro*

Preface

*T*echnology involves all phases of product development from design through delivery. Industrial practitioners must be well versed in all aspects of manufacture, including computer applications, manufacturing processes, quality control, production management, and organizational behavior, among others. Selection and processing of industrial materials are important, but students preparing for careers in industry do not need to first become proficient in materials science, as many of the good textbooks on materials available at present assume. To be successful, we feel that the authors should be specialists in materials first, but should also be well versed in product-oriented manufacturing processes, where making it right is the only practice that matters! For these reasons, this textbook is filled with case studies that illustrate industrial problems.

We also believe that teaching is important; the material in this textbook has been developed over many years of teaching industrial technology students, many of whom are now practicing in successful careers and some of whom have provided case studies that appear in the text. These students did not have rigorous mathematics backgrounds, so it was important to develop their understanding of concepts rather than their computational skills or theoretical knowledge. The thrust of this textbook is to define properties needed for applications, then relate these properties to the material properties for appropriate selection and control through processing.

It is impossible to cover all materials in a single one-semester course, yet curriculum demands do not always permit time for a two-course sequence. We have separated our treatment of materials into two volumes rather than including everything in a single cumbersome volume that might not bc fully utilized. Where only a single course is required, your emphasis can be tailored to either metals and alloys

(Volume 1) or to polymers, ceramics, and composites (Volume 2), eliminating one volume or leaving it for an elective.

There is little new information provided in these two volumes—we have borrowed freely from other sources whose permission is acknowledged with appreciation. Our contribution is in the organization of the topics and their presentation in a logical fashion to establish a basis for optimum applications of materials in manufacturing.

Volume 1 is confined to metals and alloys. We begin the first four chapters by discussing properties related to these materials and their applications. We then build a basis for understanding the selection and control needed to satisfy the demands of these materials, beginning with brief treatments of atomic structure and bonding to form crystalline solids. Defects and their consequences provide the groundwork for single phase alloys, then binary and higher order alloys. Chapter 5 describes the making of metals and alloys and the microstructure and properties resulting from solidification. Chapter 6 emphasizes the proper heat treatment to use in making the material right for its application. In Chapters 7 and 8, the important ferrous and nonferrous alloys are described, along with ideas for their use. In Chapter 9, we emphasize what happens to metals during deformation and the metalworking methods for shaping metals by deforming them. Although we emphasize proper selection and control throughout the text, we describe metal joining in Chapter 10 because many products and properties are altered by joining or during application, particularly corrosion, which is presented in Chapter 11.

We would like to thank our students for their inspiration and our colleagues at the University of Massachusetts Lowell for their encouragement and support. Particular thanks are due Professor V. E. Simms, Jr., for his invaluable discussions and assistance in locating some of the photomicrographs. We would also like to thank all those who reviewed the final manuscript: David H. Devier, Ohio Northern University; Peter Stracener, South Plains College; C. J. Law, Western New Mexico University; Bill G. Cullins, Aims Community College; Thomas F. Kilduff, Thomas Nelson Community College; and Boyd Larson, University of Wisconsin–Platteville. Finally, we could not have completed the text without the sacrifice of our families and friends, particularly our wives, Dr. Jane Dreskin and Mrs. Helen Vasilos.

Contents

3

Pure Metals and Single-Phase Alloys   35

4

Binary Alloys and Phase Diagrams   51

5

Melting and Solidification   69

6

Heat Treatment of Metals and Alloys   85

7

Ferrous Alloys   107

8

Nonferrous Metals and Alloys   131

9
Deformation Processing of Metals   153

10

Metal-Joining Processes   **185**

11

Corrosion and Corrosion Protection   **219**

1

Properties of Metals

The history of man can be measured by the development and applications of materials. It is with good reason that we refer to periods of human culture as the Stone Age, the Bronze Age, or the Iron Age because materials developments during these ages led to the production of weapons for hunting or warfare and to the production of cooking and storage utensils. Even today, our sophisticated technology is dependent on materials developments at work, at home, and at play. Enterprises such as the transportation, computer, electronics, communications, and aerospace industries are a result of our being able to study and learn about the materials needed to develop the dreams of entrepreneurs.

Our understanding of materials did not really begin until the late nineteenth century when the microscope and methods for testing materials were first developed. Up until then, materials development was purely empirical, thus limiting the technology of the time. For example, both the social and economic development of the United States was made possible by railroads. Prior to improved steelmaking by the Bessemer process, however, rails were too weak to sustain the constant travel of steam engines.

Today's materials can be classified as metals and alloys, as polymers or plastics, as ceramics, or as composites; composites, most of which are man-made, actually are combinations of different materials. Applications of these materials depend on their properties; therefore, we need to know what properties are required by the application and to be able to relate those specifications to the material. For example, a ladder must withstand a design load, the weight of a person using the ladder. However, the material property that can be measured is strength, which is affected by the load and design dimensions. Strength values must therefore be applied to determine the ladder dimensions to ensure safe use.

The properties that we will be using throughout this textbook are those of metals and alloys. They include **physical properties** such as density and melting point, **mechanical properties** such as strength and ductility, electrical and thermal properties such as conductivity, and magnetic properties. (Other properties of materials, such as optical properties, will be discussed in the second volume of this textbook on polymers, ceramics, and composites.)

The units for measurement of properties are supposedly uniform, with the International System of Units (SI units) universally acceptable. Nevertheless, conventional usage of British units in the United States has persisted in many disciplines. This mix really does not present a problem, however, because we can readily convert to SI units when measuring in British units. Table 1.1 compares the units of measurement and lists conversion factors.

Table 1.1
Measurements and material properties

Property	SI unit	British unit	Conversion factors
Length	meter	inch, foot	1 in. = 2.54 cm = 25.4 mm 1 m = 39.37 in. 1 Å = 10^{-8} cm 1 mil (.001 in.) = .0394 mm
Mass	kilogram	pound mass (lbm)	1 kg = 2.204 lbm 1 lbm = 453.7 g
Force	newton (N)	pound force (lbf)	1 lbf = 4.44 N
Stress	pascal (Pa)	lbf/in.^{2} or psi	1 Pa = 1 N/m^2 = .145 × 10^{-3} lbf/in.^{2} 1 lbf/in.^{2} = 6.89 × 10^3 Pa
Temperature	°C K (absolute)	°F °R (absolute)	°F = $\frac{9}{5}$°C + 32 K = °C + 273 °R = °F + 460

1.1   Mechanical Properties

1.1.1   Tensile Properties

Mechanical properties are always specified in material selection for structural applications. Structural design, in turn, must provide the size that is appropriate for these properties. It is typical to specify tensile properties, which simply refer to the applied forces that stretch a shape. Tensile tests are performed in universal machines such as that shown in Figure 1.1. These machines can test materials under compression, shear, and flexure. In most cases, we will specify engineering properties, which are determined from a stress-strain curve of test results.

Figure 1.1
Universal testing machine for mechanical property measurement (Courtesy of Instron Corporation.)

Engineering **stress**, σ, is defined as

$$\sigma = \frac{P}{A_o}$$

where A_o is the original cross-sectional area and P is the force that is applied. This applied force will extend or elongate the metal, causing a **strain**, ε, which is given by

$$\varepsilon = \frac{\ell - \ell_o}{\ell_o}$$

where ℓ is specimen length after force is applied and ℓ_o is the original specimen length.

When we load the specimen to failure, the stress-strain curve can be plotted, giving us useful data for specifications. Figure 1.2 shows a typical stress-strain curve for a metal sample tested to failure. Initially, there is a large *linear* increase in stress with little strain. This change is **elastic deformation** because the metal shape is completely recovered if the force is removed. The linear relation is described by the equation

$$\sigma = E\varepsilon$$

where E is the **elastic modulus,** or Young's modulus. This equation is known as Hooke's law and introduces us to the most common example of a structure independent property. No matter what we do to the material, E remains unchanged. We have to change the temperature or composition to alter it. Structure sensitive properties, on the other hand, can be altered by either heat treatment or deformation.

Figure 1.2
Typical stress-strain curve for
metal sample tested to failure

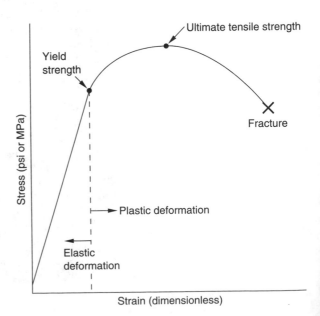

Figure 1.3
Method for determining
0.2% offset yield strength

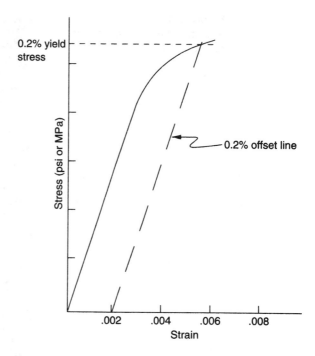

The stress-strain curve of Figure 1.2 contains a great deal of information that we will use when specifying mechanical properties of materials used in manufacturing. Most applications cannot permit any permanent deformation, much less fracture of the material, so we must design for a stress below that where **plastic deformation** first occurs. We call this stress the **yield strength** or flow strength (also referred to as yield or flow stress). The exact value of the yield strength cannot be determined with any accuracy; therefore, we use the convention denoting the 0.2% offset yield strength. This is determined by selecting the strain value of 0.002 in./in. or mm/mm (0.2%) and drawing a line parallel to the elastic or linear portion of the stress-strain curve. The intersection of this line with the stress-strain curve, illustrated in Figure 1.3, is defined as the 0.2% yield strength.

The highest stress that a metal can withstand without breaking is termed the **ultimate tensile strength**, as indicated in Figure 1.2. When the ultimate tensile strength has been reached, the sample cross section is reduced by additional loading; we call this reduction **necking**. Figure 1.4 demonstrates the necking of a mild steel sample just prior to fracture. Therefore the reduction in engineering stress beyond the ultimate tensile strength is really an artifact because we continue to divide the load by the original cross-sectional area even though the cross section has been reduced by the necking (in Chapter 9, we will examine behavior in terms of true stress and true strain). We now have identified the elastic modulus and two strengths, the yield strength used for design purposes and the ultimate tensile strength, which is the maximum strength. Both yield strength and ultimate tensile strength are structure sensitive properties that can be affected by deformation or heat treatment.

Figure 1.4
Necking of mild steel sample
during testing

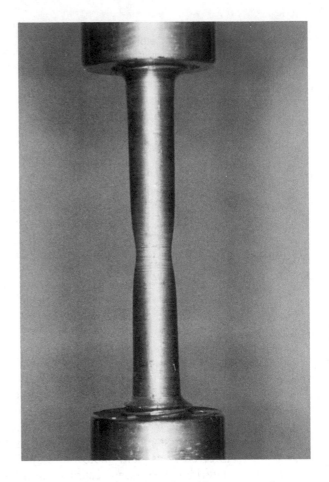

    Ductility, which is a measure of deformability without breaking, cannot be easily determined from the stress-strain diagram. The high stresses and low strains of the elastic region are experimentally measured with extensometers or strain gauges; these cannot be used for high elongation values. Also, necking that occurs is a form of ductility that is not recorded on the stress-strain diagram. We therefore express ductility in terms of the percentage elongation to fracture or the percentage **reduction in area** (% RA). These are defined as

$$\% \text{ Elongation} = \frac{\ell_f - \ell_o}{\ell_o} \times 100$$

$$\text{and } \% \text{ RA} = \frac{A_o - A_f}{A_o} \times 100$$

A standard length is marked and measured both before (ℓ_o) and after (ℓ_f) testing, and the area before (A_o) and the area in the necked region after (A_f) testing are also determined to ascertain the property values. Both % elongation to fracture and the % RA are structure sensitive properties.

Sample Problem 1.1

Steel has a modulus of elasticity, E, of 29×10^6 psi (20×10^4 MPa).

    a.   What is the stress on a steel rod 0.125 in. in diameter and 12 in. long that is stretched 0.02 in.?

    b.   What is the force causing this stress?

Solution

a. $\varepsilon = \dfrac{\ell - \ell_o}{\ell_o} = \dfrac{0.02 \text{ in.}}{12 \text{ in.}} = 1.67 \times 10^{-3}$ in./in.

$\sigma = E\varepsilon = 29 \times 10^6$ psi $\times 1.67 \times 10^{-3}$ in./in.

$= 48.43 \times 10^3$ psi

$= 48,400$ psi (333.8 MPA) *

b. $\sigma = \dfrac{P}{A}$

$48,400$ psi $= \dfrac{P}{\pi(0.125 \text{ in.})^2 / 4}$, since $A = \dfrac{\pi d^2}{4}$

$P = 48,400$ psi $\times \dfrac{\pi(0.125 \text{ in.})^2}{4}$

$= 594$ lbf (133.8 N)

Sample Problem 1.2

A yellow brass rod (Cu-30% Zn) 0.25 in. in diameter has a yield strength of 16,000 psi (110 MPa) and tensile strength of 48,000 psi (330 MPa). What is the maximum force that can be applied

    a.   without breaking the rod?

    b.   without deforming the rod permanently?

Solution

    a.   Using the ultimate tensile strength,

$$\frac{P}{A} = 48,000 \text{ psi}$$

$$P = \frac{48,000 \text{ psi}}{\pi(0.25 \text{ in.})^2 / 4}$$

$$= 2360 \text{ lb (532 N)}$$

---

* Values with more than three significant numbers cannot be supported by testing accuracy, so convention limits the answer to the rounded value of 48,400 psi.

b.   Using the yield strength,

$$\frac{P}{A} = 16,000 \text{ psi}$$

$$P = \frac{16,000 \text{ psi}}{\pi(0.25 \text{ in.})^2 / 4}$$

$$= 787 \text{ lb } (177 \text{ N})$$

1.1.2   Hardness

Standardized samples must be machined for tensile testing, which is time-consuming when we need answers for production to continue. Also, when failures occur, we cannot test small sections to determine their tensile properties. Therefore, we welcome any convenient, fast method to evaluate these mechanical properties. The most popular method is **hardness** testing, which measures the resistance of a metal to plastic deformation. In this type of test, an indentation is made on the surface of a sample with a specified load. Of course, the load, sample thickness and contour, shape of the indenter, and support for the sample all can affect the results, and we must be able to calibrate the measurement with some mechanical property. Nevertheless, this calibration has been done for us, with a number of different standard methods in daily use. In fact, hardness is frequently specified for a metal part.

The most popular hardness test is the Rockwell test, which is direct reading and can use a number of standard loads and either a spherical or conical indenter. The most common scales are the Rockwell C (150 kg load with a diamond cone indenter), which is used for hard, strong metals, and the Rockwell B (60 kg load with a hardened steel $\frac{1}{16}$-in. diameter sphere), which is used for mild steels. A Rockwell test unit appears in Figure 1.5. Hardness values have been directly correlated with the ultimate tensile strength of steel and comparative tables are readily available.

For other hardness tests, such as Brinell tests, or microhardness tests, such as Knoop and Vickers tests, that use small loads and a diamond pyramid indenter, the size of the indentation has to be measured and converted to the respective hardness scale. Microhardness tests are used for very small sample sections or for materials containing different microstructures.

In all hardness testing, it is good practice to make a number of indentations and report the average value. Following this procedure eliminates minor material variations as well as minor testing errors.

1.1.3   Toughness

Toughness is the ability of a material to absorb energy before it breaks or fractures. Consider a strong, brittle material and a weaker but ductile material, with stress-

Figure 1.5
A Rockwell hardness tester
(Courtesy of Wilson®
Instruments Division, Instron
Corporation.)

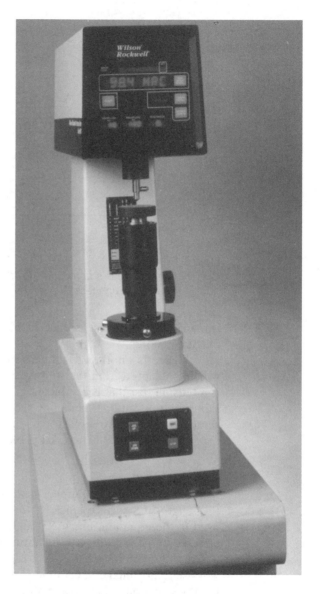

strain curves depicted in Figure 1.6. The toughness of the weaker material is higher because it is proportional to the total area under the stress-strain curve. This area represents the total energy that is absorbed in fracture.

It is common to measure the toughness of a material by an **impact** test that measures the energy absorbed in fracture under an impulse load. In this type of test, a heavy pendulum is released from a known height, the pendulum strikes and breaks the sample, then continues its upward swing. If the weight of the pendulum and the heights involved are known, the energy absorbed in fracture can be directly measured on the machine. The most common impact test for metals is the **Charpy**

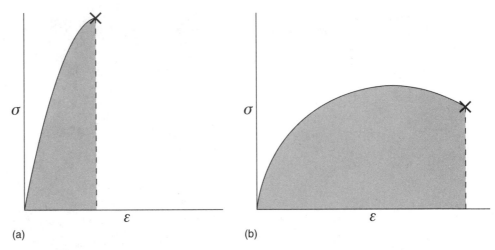

Figure 1.6
Comparison of stress-strain curves for ductile and brittle materials: (a) brittle materials, (b) ductile material. Toughness is given by the areas beneath the curves.

impact (V-notch) **test**, shown in Figure 1.7, in which the fracture begins at the machined notch.

There is another toughness measurement that we call **fracture toughness**, which is adapted from the study of fracture mechanics. The fracture toughness test is a **tension** test that is conducted in a universal testing machine using a sample containing an intentional crack, as shown in Figure 1.8. Fracture toughness is represented by K_{1c}, the stress factor to cause catastrophic failure of the sample, which is expressed in ksi in.$^{1/2}$ (where ksi is 1000 pounds per square inch):

$$K_{1c} = Y\sigma_f\sqrt{\pi a}$$

where Y is a geometric number near unity, σ_f is the stress applied at failure, and a is the length of a surface crack. If the flaw is located away from the surface, $(\frac{1}{2})a$ is used in the equation. Fracture toughness is structure sensitive and therefore can be affected by heat treatment; for example, a high-strength, low-alloy steel might have a K_{1c} value of 55 to 90 ksi in.$^{1/2}$. It is also affected by temperature, a phenomenon we will use to advantage in ceramic processing. It should be noted that K_{1c} applies only for thick sections where plane strain conditions are applicable (we will find it much more useful in the study of ceramics in volume two of this textbook).

1.1.4   *Stress Concentration*

Stress concentration is a geometric factor that can lead to failure at applied stress levels much below those anticipated for failure to occur. In brittle materials, the failure can be abrupt, caused by sudden propagation of a crack. However, the effects can be blunted somewhat by deformation preceding crack propagation for ductile

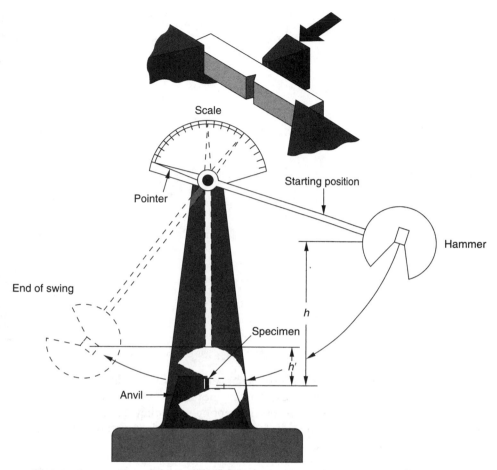

Figure 1.7
Charpy test of impact energy
(From W. Hayden, W. G. Moffatt, and J. Wulff, *The Structure and Properties of Materials*, Vol. 3: *Mechanical Behavior*, John Wiley & Sons, 1965.)

materials. Look at Figure 1.9. The highest stress at the tip of a crack is given by the equation

$$\sigma_{max}= 2\sigma \left(\frac{c}{\rho}\right)^{1/2}$$

where σ_{max} is the maximum stress at the crack tip, c is half the length of an interior crack or is the length of an exterior crack, ρ is the radius of curvature at the tip of the crack and σ is the applied tensile stress. We can readily see that the stress concentration can lead to crack propagation and localized failure when the bulk of the material is under a fairly low stress level.

Figure 1.8
Fracture toughness test

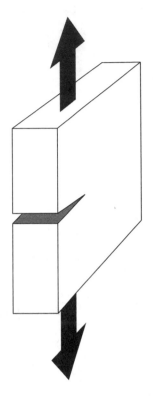

Stress concentration occurs at drilled holes, fillets, and other design configurations, but also can result from metallurgical defects. We cannot emphasize the effects of stress concentration enough, for they cannot be ignored without serious consequences. Later chapters provide several case studies in which stress concentration is demonstrated.

Sample Problem 1.3

Two ⅛-in. slots are milled into a bar 1 in. by 0.25 in. thick, reducing the width to 0.5 in. If the maximum force the bar could support without breaking was 10,000 lbf before machining, what is the maximum load it can support with the milled slots?

Solution

The ultimate tensile strength is 10,000 lbf/(1 in. × 0.25 in.), or 40,000 psi. This is the value of σ_{max}.

$$\sigma_{max} = 2\frac{P}{(0.5 \times 0.25)} \times \left(\frac{0.25}{0.0625}\right)^{1/2}$$

$$40,000\ \text{psi} = 2\frac{P}{0.125} \times 2 = 32P$$

$$\text{and } P = 1250\ \text{lbf}$$

(Without stress concentration, P would be 5000 lbf.)

Figure 1.9
Stress concentration caused
by cracks

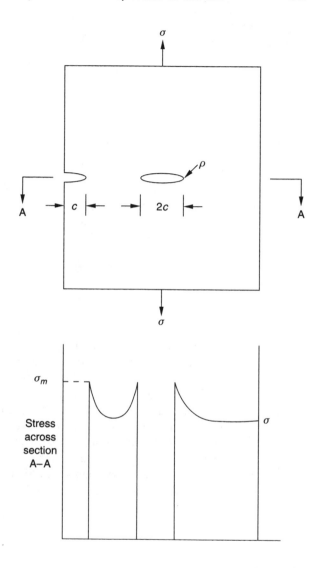

1.1.5 *Metal Fatigue*

The stress-strain curve represents failure under a single applied force. What happens when we repeatedly load a material to a level below the ultimate tensile strength? We have applied a **fatigue** load, that is, a cyclic or intermittent load. Failure under such conditions can occur at stress levels below the ultimate tensile strength after a number of cycles. As Figure 1.10 illustrates, the breaking strength for the steel under fatigue conditions approaches a value known as the **endurance limit** after many cycles. It is this endurance limit that we must design for if fatigue stress conditions are anticipated.

Figure 1.10
Fatigue failure curves

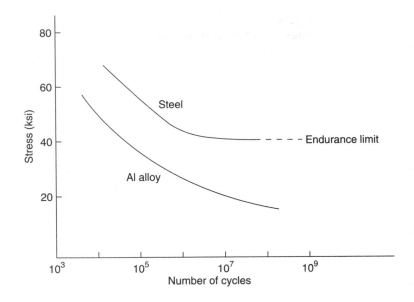

Figure 1.10 also shows that there is no definable endurance limit for aluminum. When the number of cycles continues, even for low stress values, fatigue failure can occur. A dramatic example of this type of failure is the Hawaiian Airlines incident when the cabin behind the pilots' compartment was separated from the main fuselage while in flight. Fortunately, the pilot was able to land the plane and avert a tragedy. Examination of the fractured parts showed evidence of severe high-cycle–low-stress fatigue. The failure analysis concluded that the low stresses were those of pressurization/depressurization during high-altitude flights. The number of cycles, however, was much higher than normal for the number of air miles flown because of the short flights and high altitudes required to negotiate the mountains of the islands of Hawaii. Increased inspection of the aluminum hull of an aircraft for any signs of cracking has become routine as a result of this incident.

1.1.6   Wear

We have been looking at mechanical properties that can cause failure by fracture, but there are two other ways that materials can lose their usefulness — by wear or by obsolescence. We neither can nor want to control obsolescence, but we have to understand how to prevent or protect against wear. Wear is simply the removal of small amounts of a material's surface by mechanical action. Although we usually think of wear as harmful, there are useful applications such as writing with a pencil: graphite wear particles are transferred to paper by our mechanical action.

Wear actually comprises a number of different processes that take place independently or in combination. Adhesive wear that occurs when two surfaces slide across each other is the most common, but we also encounter abrasive wear, as in the writing example and in wood refinishing with abrasive sandpaper. Less common wear phenomena are corrosive wear, fatigue wear, and deformation wear. The wear process is very complex, with many variables involved that cannot always be controlled. These variables include the mechanical properties of both surfaces, surface finish, contact pressure, lubrication, contaminants, and more.

The main transfer of force during sliding is by plowing, where asperities on one surface dig into the soft surface of another surface, forming a wear track in the softer metal. Deformation beneath the wear track leads to subsurface cracking and eventual shear to the surface, forming wear sheets that are broken into wear particles by subsequent sliding. Copper wear particles and a wear track formed in the copper by a steel slider appear in Figure 1.11.

Wear is of particular interest in manufacturing machinery whose surfaces move with respect to each other. When there is lubrication, wear does not occur because there is no metal-to-metal contact. However, there is always momentary contact during start-up or shutdown procedures. At those times, wear of the softer, bearing metal can occur. We will discuss wear resistance, such as in tool steels, in Chapter 7 and metal applications, such as bearings and bushings, in Chapter 8.

Figure 1.11
Wear particles and wear track in OFHC copper (200×)

1.2 Electrical and Thermal Properties

Whereas we specify mechanical properties for structural applications, we must specify the **electrical properties** of materials for all applications involving conduction or insulation from electrical fields. At the same time, we will be able to estimate the thermal characteristics because thermal and electrical properties are interrelated. A good electrical conductor must be a good thermal conductor and a poor electrical conductor must be a poor thermal conductor!

Conductivity, the ability to transmit electricity or heat, is inversely related to resistance, so we find it convenient to look at electrical **resistance**, given by Ohm's law:

$$R = \frac{E}{I}$$

where R is resistance in ohms, E is the voltage in volts, and I is the current in amperes. We can look at resistance of metals in the same way we looked at applied load in mechanical testing. Resistance and applied load include geometric factors and therefore do not describe the material characteristic. The material property, **resistivity**, or ρ, is a structure sensitive property that is calculated from the equation

$$R = \rho \frac{\ell}{A}$$

where ℓ is the length and A is the cross-sectional area. The common unit for ρ is the ohm-cm. Some values of resistivity are 1.7×10^{-6} ohm-cm for copper, a good conductor, 112×10^{-6} ohm-cm for Nichrome alloy, used for electrical resistance heating applications such as toaster wires, and 10^{14} ohm-cm for polyethylene, an electrical insulator.

Sample Problem 1.4

Nichrome toaster wire has a diameter (d) of 0.025 in. Would it be practical to replace this wire with copper?

Solution

$$R = \frac{\rho \ell}{A} \text{ or } \frac{R}{\ell} = \frac{\rho}{A}$$

$$\text{therefore} \left(\frac{\rho}{A}\right)_{Cu} = \left(\frac{\rho}{A}\right)_{Ni-Cr}$$

$$\frac{1.7}{d^2_{Cu}} = \frac{112}{d^2_{Ni-Cr}}$$

$$= \frac{112}{0.025^2}$$

$$d^2_{Cu} = 0.025^2 \times \frac{1.7}{112}$$

$$= 0.0000095$$

$$d_{Cu} = 0.0031 \text{ in.}$$

Such thin wire has little strength and will break easily, so replacement is impractical.

1.3 *Magnetic Properties*

Magnetic properties are critical for materials used in power generation, transformers, relays and, of course, for permanent magnets. However, the application of magnetic testing in metallurgical evaluations has largely been overlooked, perhaps because materials technologists are not familiar with these properties.

There are three types of behavior when a material is placed in a magnetic field. Most materials are **paramagnetic**, that is, they become weakly magnetized. A few, such as superconducting materials, are **diamagnetic**; they repel the applied field, giving rise to such applications as magnetic levitation for high-speed trains. Finally, we have **ferromagnetic** materials in which there is strong magnetization from the applied field. (There are also ferrimagnetic ceramics, but they are really ferromagnetic materials that depend on crystal structure for their ferromagnetic characteristics. We will learn about them in volume two.)

There are two types of ferromagnetism, which we term hard and soft. Hard magnetic materials are used for permanent magnets and soft magnetic materials are used in transformers, power generators, and recording equipment. The difference between these **hard magnets** and **soft magnets** is in the way they are magnetized. In general, application of a magnetic field, *H*, measured in oersteds, generates a higher field, *B*, measured in gauss (one gauss, however, is equal to one oersted, so the distinction is only between applied and induced field). The highest field induced is known as **magnetic saturation**, B_S, a structure insensitive property. When the applied field is removed, however, a residual field remains. This residual field, or **remanence**, B_R, is high for hard or permanent magnets and low for soft magnets. In order to remove any induced field, a field must be applied in the opposite direction. The value of this field that removes the remanence is called the **coercive force**, H_c, and is structure sensitive. The hysteresis demonstrated by application and removal of a magnetic field is shown in Figure 1.12.

We should be familiar with the following relationships:

$$B = H + 4\pi M$$
$$\text{and}$$
$$B = \mu H$$

where B is induced magnetic field, H is applied magnetic field, M is magnetization, and μ is **permeability**. Since the relationship between B and H is nonlinear, the value of permeability is not constant. For initial permeability, important for recording applications, the value of μ at 40 gauss is usually measured. This value is referred to as initial permeability, a structure sensitive property.

We learn at a very early age that permanent magnets have north and south poles and that unlike poles attract and like poles repel. Ferromagnetic materials actually contain tiny magnetic dipoles called **magnetic domains** that act in the same manner. In the demagnetized state, dipoles are continuous across a domain boundary or wall where the dipole is rotated 90° and discontinuous only across boundaries where the dipole is rotated 180°; in this latter case, dipoles are aligned, but in oppo-

Figure 1.12
Magnetization curve for
ferromagnetic material

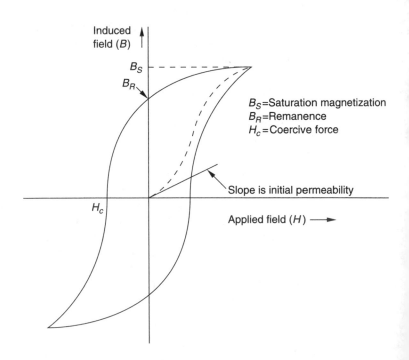

B_S=Saturation magnetization
B_R=Remanence
H_c=Coercive force

Induced field (B)

B_S

B_R

Slope is initial permeability

H_c

Applied field (H) ⟶

Figure 1.13
Domain wall movement during magnetization

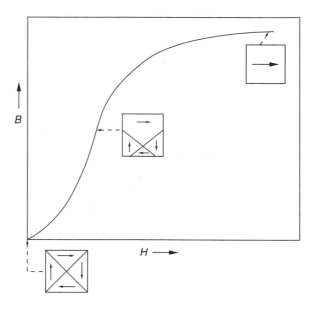

site direction. Domain walls move during magnetization, and the size of domains change until all domains are aligned in the direction of the magnetic field at B_s, the saturation magnetization. Figure 1.13 illustrates domain walls and the change in domain size during magnetization.

1.4 *Thermal Effects*

A change in temperature affects all properties to some extent. For example, ferromagnetic properties decrease as temperature is increased, disappearing at a temperature we call the **Curie temperature**, and strength decreases rapidly as the melting point is approached. Electrical conductivity increases as temperature decreases and completely disappears in certain materials we call superconductors. The most common effect of temperature change, however, is thermal expansion and contraction, which can lead to mechanical stress levels in materials subjected to changing temperature. In some instances, if thermal conductivity is low and thermal changes are high, failure can occur by thermal shock.

The strain caused by thermal expansion is represented by

$$\varepsilon = \frac{\Delta \ell}{\ell_o} = \alpha \Delta T$$

where α is the thermal expansion coefficient and ΔT is the change in temperature. Strain can be translated into stress if the object is constrained from expansion by using Hooke's law.

Sample Problem 1.5

The thermal expansion coefficient of steel is $17 \times 10^{-6}/°C$. The temperature of a steel rod 2 ft (60.96 cm) long, fixed at both ends, is cooled from 20°C to 0°C. Assuming no initial stress,

    a.   is the rod under tension or compression at 0°C?

    b.   what is the value of the stress at 0°C?

Solution

    a.   The rod contracts, but the ends are fixed, causing the rod to be in tension.

    b. $\Delta T = 20°C$, $\sigma = E\alpha\Delta T$

$$= 29 \times 10^{6} \text{ psi (0.2 MPa)} \times 17 \times 10^{-6}/°C \times 20°C$$

$$= 9860 \text{ psi (68 MPa)}$$

Summary

We can only understand materials by first understanding their properties and how the properties limit applications. This introductory chapter has been limited to those properties we must be familiar with. The most frequently specified properties are the mechanical properties, particularly the tensile properties that require knowledge of stress, strain, and strength. Closely related to strength is hardness, which is more conveniently measured in many instances. Toughness reflects energy absorption in breaking and is usually measured by impact testing. Perhaps the most important concept for structural applications is the effect of stress concentration, which causes failure to occur at otherwise reasonable stress. Other properties that we must be concerned with are electrical or thermal properties and magnetic properties. Resistivity is the single most important electrical or thermal property; it is structure sensitive, that is, it is affected by heat treating or deformation of the metal. There are three types of material behavior in a magnetic field: diamagnetism, paramagnetism, and ferromagnetism. Not only are the magnetic properties necessary for some applications, but we can learn about behavior of materials, particularly ferromagnetic metals, by the changes in their magnetic properties.

Terms to Remember

Charpy impact test	magnetic saturation
coercive force	mechanical properties
compression	necking
conductivity	paramagnetic
Curie temperature	permeability
diamagnetic	physical properties
ductility	plastic deformation
elastic deformation	reduction in area (RA)
elastic modulus	resistance
electrical properties	resistivity
elongation	remanence
endurance limit	soft magnet
fatigue	strain
ferromagnetic	stress
fracture toughness, K_{1c}	stress concentration
hard magnet	tension
hardness	toughness
impact	ultimate tensile strength
magnetic domains	yield strength
magnetic properties	

Problems

1. Describe the shape of the engineering stress-strain curve, distinguishing between elastic and plastic deformation. What happens if the load is removed in these regions?

2. A standard steel sample with a diameter of 0.505 in. is tested to failure in tension. The maximum load was 6400 lbf, the 2.00-in. gauge length was 2.644 in. after testing, and the necked diameter was 0.476 in. after testing. Calculate the ultimate tensile strength, the % elongation to fracture, and the % reduction in area in both British units and SI units.

3. Describe how to determine yield strength, ultimate tensile strength, and the elastic modulus.

4. A 10-mm diameter steel rod is loaded elastically with a force of 5000 N. Calculate the stress and strain in both British and SI units.

5. Describe why you might use hardness testing instead of tensile testing for incoming inspection of materials.

6. A steel wire 0.025 in. (0.0635 cm) in diameter is stretched elastically by a force of 30 lbf (133 N). If the temperature is raised by 20°C, how must the load be changed to maintain the same length of the wire?

7. For resistance heating, heat transfer is a function of the surface area of the heating element. Compare the effectiveness of heat transferred from wires having the same resistance and length. Assume that one wire is Nichrome and the other a copper-nickel alloy with half the resistivity of Nichrome. Which wire gets hotter?

8. Draw the magnetization curve for
    a. a diamagnetic metal
    b. a paramagnetic metal
    c. a ferromagnetic metal.

9. For a ferromagnetic metal, define the saturation magnetization, permeability, coercive force, and remanence. Which of these can be altered by processing?

10. Draw the magnetic domain structure of a ferromagnetic material
    a. in the demagnetized state
    b. in a partially magnetized state
    c. at saturation magnetization.

11. Compare the stress-strain curves of two 0.505-in. diameter steel rods, the first solid and the second with a V-notch cut into it, reducing its diameter to 0.475 in. (assume $\rho = 0.05$ in.).

12. Explain the conditions that require you to design with the endurance limit in mind.

13. Examine the materials in an incandescent light bulb. List each different component, what its purpose might be, and what material properties must be considered in its selection.

14. Consider the designs of a mountain bike and a racing bike. What factors will affect your selection of the metal frame for each?

2

Atoms, Bonding, and Crystal Structures

In Chapter 1, we looked only at properties of metals and alloys because it is necessary to understand the relationships among their properties, selection, and processing. In this chapter, we begin with an examination of the base upon which we will build — atoms that are bonded together in order to make up a solid structure. Each atom of a particular element is identical to every other atom of that element but unlike an atom of any other element. An **atom** consists of a nucleus, made up of positively charged **protons** plus uncharged **neutrons**, and negatively charged **electrons** that are in orbit around the nucleus, much like the planets orbit around the sun. The number of protons and electrons must be identical for each element because an atom by itself is electrically neutral. The weight of a proton or neutron is very heavy compared to an electron, so the weight of an atom is dependent only on the number of protons and neutrons in its nucleus. The size of the atom, however, is affected by its number of electrons. Each electron orbit, or shell, can have only a certain number of electrons; additional electrons must be in ever-larger diameter orbits (reasons for this are of interest to nuclear physicists, but not to us).

The elements have been arranged into the **periodic table** of the elements, according to their different characteristics. This periodic table, reproduced inside the front cover, arranges the elements in order of increasing **atomic number**, which is the total number of electrons for an element. Each successive row of the table represents an additional electron shell. The columns represent the **valence** number of the element, that is, the number of electrons in the outer shell. We will find that this valence shell is the most important characteristic because it determines how elements can combine with each other.

There are few properties that can be related only to atomic structure. Perhaps the most familiar property is ferromagnetism, which is attributed to an unfilled inner electron shell and occurs only for iron, cobalt, and nickel. Atomic physicists tell us that the color of an element is also characteristic of the atomic structure.

2.1 Atomic Bonding

There are two types of **atomic bonding**, that is, bonds that can join atoms together — primary bonds, which are strong, and secondary bonds, which are comparatively weak. We will learn only about the primary bonds, but must remember the role of secondary bonds when considering overall bulk strength. It is easy to recognize how primary bonds are formed when we look at the inert elements, all gases, in the periodic table of the elements. Helium, neon, argon, krypton, xenon, and radon do not combine with other elements because their valence shells are completely full. There are eight electrons in a filled shell, except in the case of the innermost shell, which is filled by two electrons. If we can join atoms so that their outer shells are filled, then the bond should be strong. This is the case in a **covalent bond**, which is the strongest of the primary bonds. Valence electrons are shared between nearest neighbor atoms to achieve the eight-electron configuration that fills the outermost electron shell. Bonding of fluorine atoms to form a fluorine molecule, depicted in Figure 2.1, is an example of a covalent bond.

A very important example of a covalent bond is that of carbon, with four electrons available for bonding; a carbon atom can form four covalent bonds with four adjacent carbon atoms. Each carbon atom provides one electron to make up the electron pair needed for a covalent bond. The four pairs constitute an octet of electrons. This carbon bond is strong, directed, and rigid, as exemplified by the properties of diamond, which is the crystalline form of covalent carbon.

Other materials have covalent bonds but also have secondary bonds present. For example, graphite forms as a structure that has layers of carbon atoms that are covalently bonded, but between layers there is only secondary bonding. Polymers are also covalent bonded long-chain structures that have weak secondary bonds between the chains.

The inert outer electron shell of an atom can also be formed by an **ionic bond** (or, electrovalent bond). Groups such as the halogens can accept an electron, so they have −1 valence; groups such as the alkali metals can give up an electron, thus they have +1 valence. The inert ions of unlike sign will attract each other and those

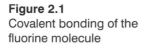

Figure 2.1
Covalent bonding of the fluorine molecule

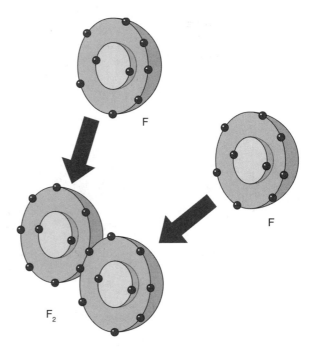

of like sign will repel each other because of the forces of **electrostatic attraction** and **electrostatic repulsion**, respectively. The formation of beryllium oxide, an important ceramic material, is shown schematically in Figure 2.2. Ionic bonds are not unidirectional like the covalent bonds, and forces attract as many ions of opposite charge as can touch at the same time. For example, in sodium chloride, NaCl, which is common table salt, a sodium ion will attract as many chlorine ions as will fit around it. Bonding is directly related to the geometrical factors as well as the need to preserve electrical neutrality of the solid.

In contrast to other major material groupings — that is, plastics where the primary bonding between atoms is covalent and salts and ceramics where the bonding can be ionic or covalent — metals owe their behavior to the unique features of the metallic bond. The **metallic bond** is somewhat similar to the covalent bond in which outer electrons are shared. Aluminum metal, for example, with three valence electrons per neutral atom requires five additional electrons to satisfy the eight valence electrons required for bond stability. In normal covalent bonding, where each linked atom contributes one member of a shared electron pair, an aluminum atom can form only three bonds, acquire only three nearest neighbors, and add to its valence orbit only three additional electrons. This, however, does not represent a full valence shell of eight. Thus there is no way in which aluminum atoms can combine by normal covalent bonding to provide a stable closed shell of eight valence electrons. A metal atom, however, can acquire as many additional valence electrons as needed to complete a stable outer shell by electron sharing on a grand scale. This involves distant atoms as well as those closest to it. Since the metallic and the cova-

Figure 2.2
Ionic bonding of
beryllium oxide

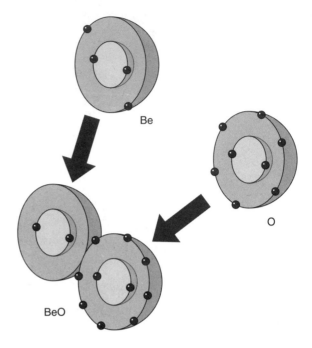

lent bond both involve the sharing of valence electrons, we often describe the metallic bond as a *mobile* covalent bond.

The free electrons that characterize the metallic bond promote good electrical and thermal conductivity of all metallic solids. This trait is well known and metals are distinguished from nonmetals by the inset added to the periodic table of the elements shown in Figure 2.3. (The inset also identifies the crystal structure of metallic elements, which we will be looking at shortly.) Because electrons scatter light, metals that have these mobile electrons are all opaque, in contrast to some transparent and translucent materials with covalent or ionic bonds.

2.2 *Crystal Structure*

In metals and alloys, the solid state is characterized by a **crystal structure**, a long-range three-dimensional order of atoms that satisfies the metallic bond requirements. The number of nearest neighbor atoms, called the **coordination number**, is high and the most common structures are those in which the atoms are packed together as closely as mutual repulsion of their electronic and nuclear charges permits.

With some minor exceptions, all solid metals are crystalline, a fact that is strongly reflected in their properties and behavior. The crystal **lattice** is a geometric array of points used to describe three-dimensional space; each point has identical surroundings. The building block for the crystal lattice is the **unit cell**, the smallest

Key

Key: Non-metallic ▨

Metallic crystal structures:
bcc—body-centered cubic
fcc—face-centered cubic
hcp—hexagonal close-packed
nch—not cubic or hexagonal
p—polymorphic (crystal structure changes with temperature)
l—liquid at room temperature

Crystal structure table (Groups IA–0):

IA	IIA											IIIA	IVA	VA	VIA	VIIA	0
	hcp																
bcc	hcp	IIIB	IVB	VB	VIB	VIIB		VIII		IB	IIB						
bcc	hcp	p	p	bcc	bcc	bcc	p	p	fcc	fcc	hcp			fcc			
bcc	p	p	p	bcc	bcc	bcc	p	fcc	fcc	fcc	hcp		nch	hcp	nch		
bcc	p	p	p	bcc	bcc	bcc	p	fcc	fcc	fcc	hcp		d	nch	p		
bcc	bcc	p						p	fcc	fcc	l		p	nch	p		

Periodic table of the elements (atomic number, symbol, atomic weight):

IA	IIA	IIIB	IVB	VB	VIB	VIIB	VIII			IB	IIB	IIIA	IVA	VA	VIA	VIIA	0
1 H 1.008																	2 He 4.003
3 Li 6.941	4 Be 0.012											5 B 10.81	6 C 12.01	7 N 14.01	8 O 16.00	9 F 19.00	10 Ne 20.18
11 Na 22.99	12 Mg 24.31											13 Al 26.98	14 Si 28.09	15 P 30.97	16 S 32.06	17 Cl 35.45	18 Ar 39.95
19 K 39.10	20 Ca 40.08	21 Sc 44.96	22 Ti 47.09	23 V 50.94	24 Cr 52.00	25 Mn 54.94	26 Fe 55.85	27 Co 58.93	28 Ni 58.71	29 Cu 63.55	30 Zn 65.38	31 Ga 69.72	32 Ge 72.59	33 As 74.92	34 Se 78.96	35 Br 79.90	36 Kr 83.80
37 Rb 85.47	38 Sr 87.62	39 Y 88.91	40 Zr 91.22	41 Nb 92.91	42 Mo 95.94	43 Tc 98.91	44 Ru 101.07	45 Rh 102.91	46 Pd 106.4	47 Ag 107.87	48 Cd 112.4	49 In 114.82	50 Sn 118.69	51 Sb 121.75	52 Te 127.60	53 I 126.90	54 Xe 131.30
55 Cs 132.91	56 Ba 137.33	57 La* 138.91	72 Hf 178.49	73 Ta 180.95	74 W 183.85	75 Re 186.2	76 Os 190.2	77 Ir 192.22	78 Pt 195.09	79 Au 196.97	80 Hg 200.59	81 Tl 204.37	82 Pb 207.2	83 Bi 208.98	84 Po (210)	85 At (210)	86 Rn (222)
89 Ac** (227)																	

* Elements 57–71 are lanthanide series or rare earths.
** Elements with atomic number greater than 89 are actinide series, which includes Th and U.

Figure 2.3
Periodic table of the elements

27

Figure 2.4
The cubic lattice of positions that define space. The unit cell is the basic building block for the lattice.

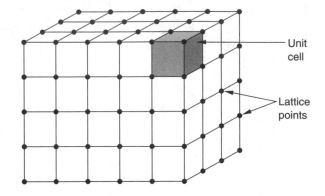

Unit cell

Lattice points

subdivision of the lattice that still retains the overall characteristics of the entire lattice. The relationship of crystal lattice and unit cell is shown in Figure 2.4. Although there are seven different crystal systems, we will be concerned only with the cubic and hexagonal in this textbook.

Principal crystal structures in metals are the **body-centered cubic (bcc), face-centered cubic (fcc),** and **hexagonal close-packed (hcp)**. These structures are illustrated in Figure 2.5, and specific elements that have these structures are identified in the inset of Figure 2.3. All unit cells have atoms at each corner; the bcc has an additional atom in the center of the cube and the fcc has atoms in the center of each face. Each corner atom is shared by eight unit cells, however, so only one eighth of the atom belongs to a specific unit cell. The body-centered atom belongs wholly to the bcc unit cell. Thus, we have $(8 \times \frac{1}{8}) + 1 = 2$ atoms per unit cell for bcc structures. There are six face-centered atoms in the fcc structure, but each is shared by two unit cells, so we have $(8 \times \frac{1}{8}) + (6 \times \frac{1}{2}) = 4$ atoms per unit cell for fcc structures.

Certain elements, notably iron, are identified as polymorphic in Figure 2.3. This term refers to a change in the crystal structure as a function of temperature below the melting point. Iron at temperatures below 912°C is bcc, but between 912°C and 1394°C it is fcc. At 1394°C it transforms again to the bcc structure that is stable up to the melting point. We will find this feature very important in heat-treating steel, discussed in Chapter 6.

The side of a unit cubic cell is called the lattice parameter, a, which can be described in terms of the atomic radius. Additional lattice parameters are required to define the size and shape of noncubic cells. Hexagonal unit cells also require two dimensions, a and c, that are perpendicular to each other; the angle between the three a axes is 120°. It is useful to consider structure-property relationships at this level because of the obvious importance to potential applications.

If we examine the bcc structure, we can see that the densest packing occurs when atoms touch along the body diagonal, designated AB in Figure 2.6. Using simple geometric relationships, the length of the body diagonal is $a \times 3^{1/2}$. (Directions within the unit cell and planes in the unit cell can be described mathematically but will not be covered in this text.) If we examine the face of the bcc structure in the

Figure 2.5
Main crystal structures of metals:
(a) face-centered cubic (fcc),
(b) body-centered cubic (bcc),
and (c) hexagonal close-
packed (hcp)

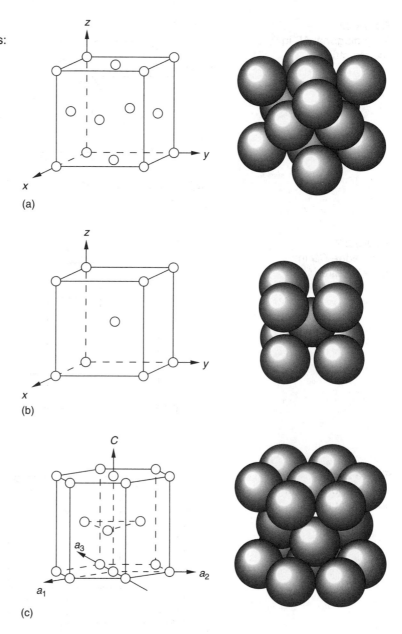

(a)

(b)

(c)

same manner, atoms do not touch along either the cube edge, designated *CD* in Figure 2.7, or along the face diagonal, designated *EF* in Figure 2.8. The significance of this positioning is that observed properties in single crystals are dependent on crystallographic direction. In the case of bcc iron, the elastic modulus measured along the body diagonal (*AB*) of a crystal is 40×10^6 psi whereas that along a cube edge (*CD*) is only 20×10^6 psi.

Figure 2.6
Body diagonal *AB* in
bcc structure

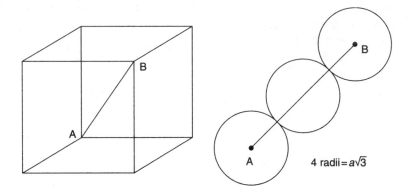

4 radii = $a\sqrt{3}$

Figure 2.7
Cube edge *CD* in
bcc structure

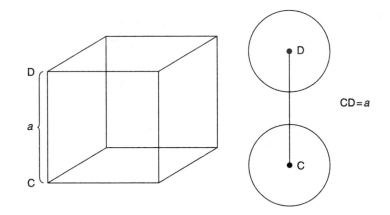

$CD = a$

Figure 2.8
Face diagonal *EF* in
bcc structure

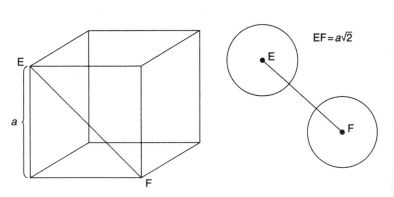

$EF = a\sqrt{2}$

Figure 2.9
Cube face diagonal *AB* in
fcc structure

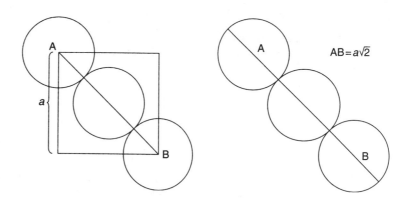

The elastic modulus for a material is a measure of intrinsic bond strength along a particular direction, and maximum strength for metals is usually observed with highest atom concentration per unit distance that is associated with maximum bonding. However, an iron nail will have an elastic modulus of about 30×10^6 psi because it is polycrystalline and there is random crystal orientation along the length of the nail. We will see in Chapter 9 that deformation processing can produce texture or preferred orientation that we can take advantage of for certain applications.

For fcc crystals, the atoms touch along the face diagonal, as shown in Figure 2.9, and other crystal directions are more loosely packed. In hcp metals, atoms touch in the a_1, a_2, and a_3 basal directions, as shown in Figure 2.10. (It is sometimes useful to describe the *linear density*, i.e., the fraction of the length of any direction occupied by atoms.)

Figure 2.10
Basal plane of hcp crystals

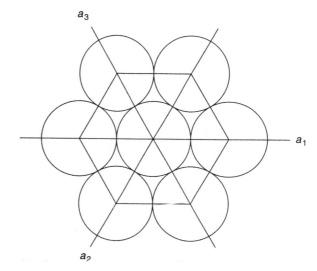

Figure 2.11
Example of equivalent planes with densest atomic packing in bcc structure

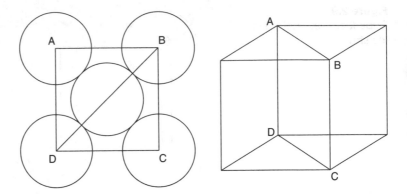

We are also interested in planar density because the smoothest planes are going to be those that are the most densely packed. We will find this to be important as well in deformation of metals in Chapter 9. Of course, the basal plane is the densest packed plane in hcp crystals, but in bcc crystals the densest packed plane is the diagonal plane, as shown in Figure 2.11, and in fcc crystals it is the plane defined by the three intersecting closest packed directions, as shown in Figure 2.12.

Sample Problem 2.1

Density of Pure Metals

Compare the atomic density of iron at 911°C and at 913°C.

Solution

At 911°C, iron is bcc. Atoms touch along the body diagonal, so $4r = a \times 3^{1/2}$, or $a = 4r/3^{1/2}$ where r is the radius of an iron atom. The cube (unit cell) volume is a^3, or $(4r/3^{1/2})^3$. The volume of a sphere (atom) is $(4\pi r^3)/3$ and there are two atoms per unit cell. Thus,

$$\text{atomic density} = 2 \times \frac{(4\pi r^3)/3}{(4r/3^{1/2})^3} = 0.68$$

At 913°C, iron is fcc. Atoms touch along the face diagonal, so $4r = a \times 2^{1/2}$, or $a = 4r/2^{1/2}$. The cube volume is a^3, or $(4r/2^{1/2})^3$. The volume of a sphere is $(4\pi r^3)/3$ and there are four atoms per unit cell. Thus,

$$\text{atomic density} = 4 \times \frac{(4\pi r^3)/3}{(4r/2^{1/2})^3} = 0.74$$

(This density difference is important in heat-treating steel, an iron alloy, because it will contract on heating through the transformation and expand upon cooling.)

Figure 2.12
Example of equivalent
plates with densest atomic
packing in fcc crystals

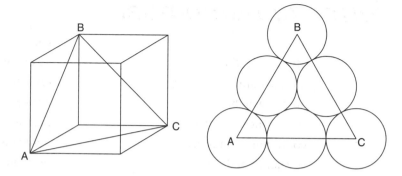

The sample problem points out how a physical property like density depends on crystal structure. If we know the atomic radius and we know the number of atoms in the unit cell, we can determine the lattice parameter from these geometric relationships. Density is weight per unit volume and the weight of one atom in grams is the **atomic weight** divided by Avogadro's number, 6.023×10^{23}. Most properties of interest for industrial applications, however, will depend on other characteristics that we have yet to study.

Summary

The basic building block that we use to develop our structure-property relationships is the elemental atom, made up of neutrons, protons, and electrons. Elements are organized in a periodic chart that points to similar behavior based on valence. How atoms bond together to form a solid is very important. The strongest bond, the covalent bond in which electrons are shared by nearest neighbors, can be rigid. Polymers and some ceramics have covalent bonds. Ionic bonds form because of the electrostatic attraction forces on unlike positive and negative ions; these bonds are also formed between nearest neighbors. Metallic bonds are more flexible and occur by the sharing of outer electrons by adjacent and distant atoms, a mobile covalent bond. Because of the free electrons, metallic elements are good electrical and thermal conductors and are opaque because the electrons scatter light. On the other hand, those ceramics and polymers that have strong covalent bonds make good electrical and thermal insulators.

All metals and alloys are crystalline solids, that is, they have atoms located in specific positions in long-range three-dimensional lattices. Most metals of industrial interest are cubic or hexagonal, with both face-centered cubic and body-centered cubic crystal structures being common as well as the hexagonal close-packed structure. Certain physical properties, such as density, are attributable to the crystal structure. Geometric relationships such as densest packed directions and planes will prove to be important characteristics that affect processing.

Terms to Remember

atom

atomic bonding

atomic number

atomic weight

body-centered cubic (bcc)

coordination number

covalent bond

crystal structure

electrons

electrostatic attraction

electrostatic repulsion

face-centered cubic (fcc)

hexagonal close-packed (hcp)

ionic bond

lattice

metallic bond

neutrons

periodic table

protons

unit cell

valence

Problems

1. Explain what valence is.
2. Describe in your own words the difference between covalent bonding and metallic bonding.
3. Based on the description of ionic bonds and nearest neighbor relationships, draw a unit cell showing the location of sodium and chlorine ions in the sodium chloride cubic lattice.
4. Describe in your own words what a crystalline solid is.
5. What are the room temperature crystal structures for the following metals?
    a. Fe
    b. Cu
    c. Ag
    d. Au
    e. Mo
    f. Nb
    g. W
    h. Al
    i. Mg
    j. Ni
6. Compare the linear density for the face edge of a bcc crystal with the linear density for the face edge of an fcc crystal.
7. Compare the planar density for the face of a bcc crystal with the planar density for the face of an fcc crystal.

3

Pure Metals and Single-Phase Alloys

Pure metals are the exception rather than the norm for manufacturing. Although some of the exceptions are commercially important, such as copper for electrical wiring and silicon in the semiconductor industry, they are more or less isolated from standard manufacturing environments. In this chapter, we look at pure metals, the types of defects that can be present in their crystal structure, and what the effects of these defects are on properties. Then we consider additions made intentionally to the pure metals, thus forming **alloys**, and examine why they are added and what effects they have on properties.

When we refer to purity, we normally associate it with chemical purity. Materials of commercial importance are derived by chemical reduction of ores, processes that add expense to the metal extracted. It follows that the more pure the metal is, the more extensive the extraction process, and the more expensive the metal. Therefore, we will find only rare applications of pure metals because of the economic factor. Technically, we are more interested in the effects of defects and impurities on the properties of the metals.

3.1   Characteristics of Pure Metals

A chemically pure metal is never perfect; there are physical defects that can never be eliminated but can be controlled. All crystal structures contain these defects, some of which are beneficial for processing (as we will see in Chapter 9). If we were to have a perfect crystal of any metal, however, what would its characteristics be? Well, its strength would be greater but would depend on the direction of the forces on the crystal. (If properties of a material are the same no matter what the direction of measurement might be, we call it *isotropic*. To distinguish property dependence on direction in the crystal, therefore, we use the term *anisotropy*.) The electrical and thermal conductivity of a pure metal would be the highest possible (although copper would still be superior to elements like iron), and all other properties would also be better than those of impure metals. But we don't want to dwell on a perfect world that we will never see, so let's look at the types of defects that occur and how they influence our materials.

3.2   Defects in Materials

We can describe **defects** geometrically because they are physical imperfections. Thus we will talk of point defects, linear defects, and volume defects. All of these occur and all can be important in structure-property relationships.

Single-dimensional point defects can occur at a crystalline lattice site or in the interatomic spaces in the crystal. Although there are other point defects, the only ones that will concern us are **vacancy**, where an atom is missing from its lattice site, and **impurity**, where a foreign atom can replace an atom of the pure metal (a substitutional impurity) or can occupy a hole, or interstice, in the lattice (an **interstitial** impurity). Figure 3.1 illustrates these types of defects. Later in this chapter, we will see that alloys are based on the substitution or interstitial position of alloy atoms.

Figure 3.1
Point imperfections in
an atomic plane

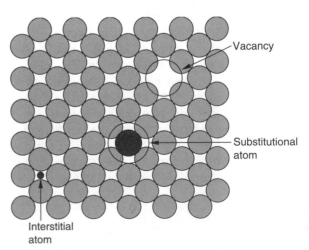

Vacancy

Substitutional
atom

Interstitial
atom

What effects do point defects have on properties? Vacancies are actually helpful because they make it easier to move atoms around in the lattice, a process we call *diffusion*, which is the basis for heat-treating metals and alloys. The situation is different with impurities, however, because properties are often adversely affected. Electrical resistivity is always increased by impurities, although the effect is dependent on what the impurity is, as shown in Figure 3.2. Even in alloys, impurities have many deleterious effects. Figure 3.3 illustrates the influence of sulfur on the initial magnetic permeability of commercial Fe-49% Ni alloys. A special impurity effect is termed *hydrogen embrittlement*, where all ductility is lost because of a few parts per million (ppm) of hydrogen. Hydrogen atoms are the smallest atoms and fit into the many interstices of bcc metals. Although the atomic packing factor for fcc metals is higher than that for bcc metals, the number of interstices is smaller. The size of the interstices is larger, however, so hydrogen escapes from the lattice easily and fcc metals are not embrittled by it. In bcc metals, on the other hand, hydrogen atoms embrittle the metal even for very low concentrations of a few parts per million (see Case Study 3.1).

Advances in steelmaking have improved properties through impurity control. For example, Figure 3.4 shows in chart form the increased ductility and toughness of a high-strength, low-alloy steel casting. The improved properties are attributable to the reduction of sulfur from 170 ppm to 20 ppm. The increase in % RA and the Charpy impact strength is almost 50% with no loss in strength.

Figure 3.2
Influence of selected impurities on electrical resistivity of copper

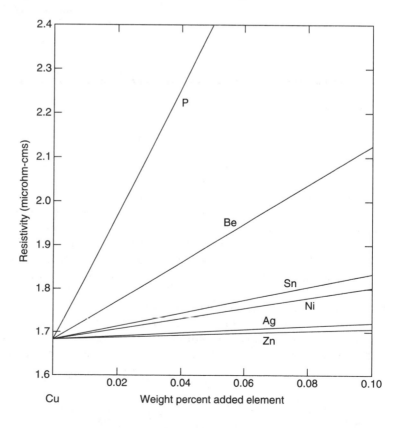

Figure 3.3
Influence of sulfur on magnetic permeability of Fe-49% Ni alloys

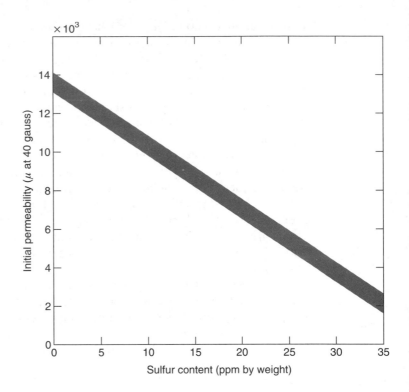

Case Study 3.1

How Vendor Change Can Affect Manufacturing

AMB Corporation manufactures cabinets that house electronic equipment. The cabinets are made of channel iron, angle iron, and shelves that can be altered to fit many different configurational needs. The parts are assembled by using small self-tapping screws, shown in Figure 3.5. The steel screws have traditionally been sent to a local electroplating company, GalvPlate, Inc., for zinc plating to provide corrosion resistance (see Chapter 11). A salesman for a new electroplating company, EPC Corporation, approached AMB management and convinced them that EPC could provide zinc-plated screws more efficiently and economically, in line with the new Just-In-Time manufacturing program at AMB.

When the EPC screws were introduced to the production line, productivity was immediately affected adversely. Screws broke when inserted using screw guns set to the standard torque. The broken screws showed no ductility, so samples were sent out for failure analysis. Metallography indicated no anomalies in microstructure and the hardness was in conformance with specifications. However, scanning electron microscopy showed numer-

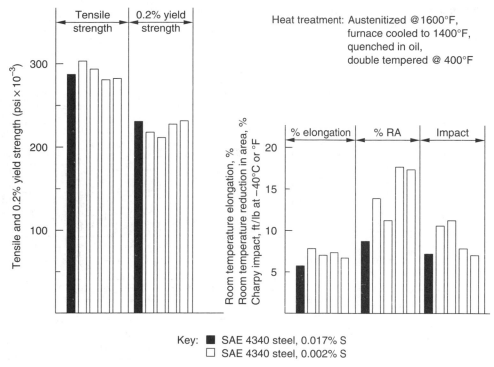

Figure 3.4
Effect of sulfur content on the ductility and toughness of high-strength steel castings

ous cracks and brittle, intergranular fracture characteristics. Examination at very high magnification revealed small bubbles, shown in Figure 3.6, that are characteristic of hydrogen embrittlement. Chemical analysis showed no unusual deviations from specified chemistry, but the hydrogen content was 14 ppm. These results were enough to require immediate action. All EPC screws were withdrawn from production and purchasing contracts were reopened with GalvPlate to again supply the zinc plating, fortunately in time to fit into the JIT program.

Two-dimensional or linear defects that occur in materials are known as **dislocations**, descriptive of the disturbance of the crystal lattice. Although dislocations are complicated and differ for different crystal structures, they are only combinations of two simple dislocations — the **edge dislocation**, which is formed by inserting an extra plane of atoms into the lattice, and the **screw dislocation**, which is formed by a shear of the lattice. These two simple dislocations are shown in Figure 3.7.

Figure 3.5
Self-tapping screw for
electronic cabinets

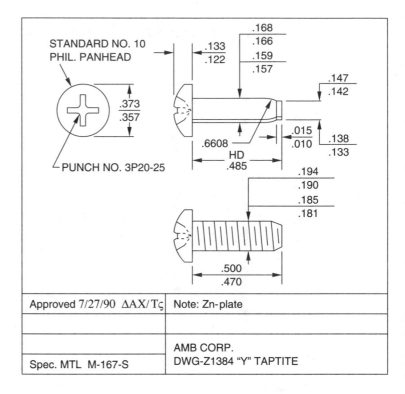

The effect of dislocations on properties is as complicated as the dislocations themselves. As we will learn in Chapter 9, dislocations permit easier deformation of metals by moving through the lattice, so if we stop them from moving we will effectively make the metal stronger. Halting the dislocation is the primary strengthening method for metals, through heat treatment, cold work, alloying, and so on. On the other hand, dislocations reduce properties such as electrical conductivity and initial magnetic permeability. Therefore, we must be aware of processing that affects the properties of our materials.

Volume defects include voids (which are formed by coalescence of vacancies or can be the result of shrinkage or gas evolution in solidification) and surface defects such as the **grain boundary** that separates single crystal grains of different orientation in polycrystalline materials. Voids are to be eliminated wherever possible because they represent stress raisers and can cause unplanned failures in applications where mechanical strength is critical.

Surface defects, on the other hand, cannot be avoided, and we do not necessarily want to eliminate them. In Chapter 9, we will see that strength can be related to the **grain size** of polycrystalline metals. Grain boundaries are the imperfect surfaces that separate the many single crystal grains in these metals. Because of their imperfect nature, grain boundaries react differently to chemicals; we can examine

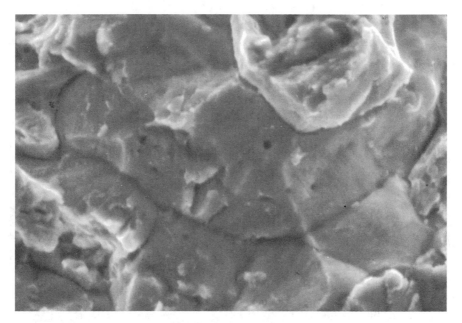

Figure 3.6
Scanning electron microscope (SEM) micrograph of intergranular fracture of hydrogen embrittled steel, with hydrogen bubbles visible on surfaces of grains (3000×)

polished and etched metals in a microscope because the grains and grain boundaries reflect differently. Much of our understanding of metals has been derived by the study of their microstructures.

Although it is the grain size that can be related to the average diameter, it is conventional to refer to the grain size number as defined by the American Society for Testing and Materials (ASTM). The ASTM grain size number is determined by the formula

$$N = 2^{n-1}$$

where N is the number of grains per square inch at 100× and n is the ASTM grain size. The ASTM grain size number is frequently specified, but we should not be confused that the value of n is higher for smaller grains, because this is caused only by the definition of the grain size number. We usually determine the grain size by comparison to published charts. Figure 3.8 illustrates some specific grain sizes.

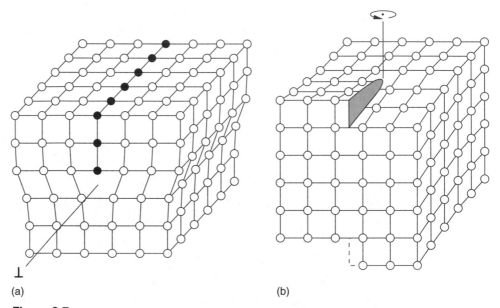

(a)   (b)

Figure 3.7
Geometry of simple dislocations: (a) edge dislocation and (b) screw dislocation. The lines normally used to represent the dislocations are also indicated, as are the symbols for them, ⊥ and ⊙.

Another surface defect is called *twinning* and is caused by two orientations within a grain that are mirror reflections of each other. Twins usually occur only in fcc metals that have been annealed, such as the cartridge brass shown in Figure 3.9.

3.3 Single-Phase Alloys

The alloys that we will examine are metallic materials that combine two or more different elements. We consider additions to be alloying elements rather than impurities when the effect of the second, added, element is substantial. In this section, we are going to look at single-phase alloys that are solutions not terribly different from a brine solution of salt and water. If we add one element to another, we call the added element the *solute* and the element being diluted the *solvent*.

In a **solid solution** (very simply, a solution made up of solids), the position of solute atoms in the solvent lattice can be substitutional or interstitial, much the same as the impurity positions given in Figure 3.1. The properties will change as we increase the solute concentration. For example, the color of copper changes from

Figure 3.8
Comparison of standard ASTM grain sizes of low-carbon steel (100×): (a) nominal ASTM grain size No. 2, (b) nominal ASTM grain size No. 4, (c) nominal ASTM grain size No. 6, (d) nominal ASTM grain size No. 8

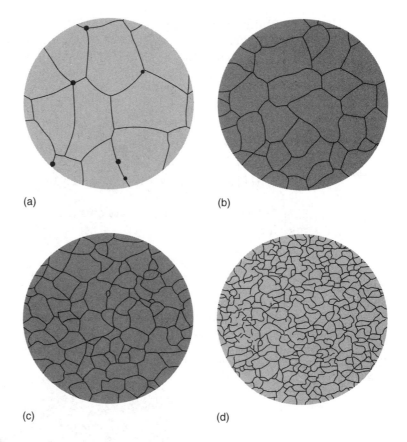

(a)　　　　　　　(b)

(c)　　　　　　　(d)

red to yellow as we add zinc to make brass. In solid solutions, we also can reach a solubility limit that causes precipitation of a second solid phase, much the same as sugar precipitates from a cup of coffee when we put in too much or when sweetened coffee cools. What controls the solubility and how the properties are changed below the solubility limits are important questions to be answered.

Atomic size is perhaps the most important characteristic of an alloy element. The only atom that is important in metal alloys and is small enough to occupy an interstitial position is carbon. Steel is an alloy of iron and carbon. Solubility of carbon in the bcc low-temperature iron, however, is small in comparison to solubility in the fcc high-temperature iron. This difference is a result of the size of the interstices that the carbon atoms fit into. In fcc, the atomic packing factor is higher than in bcc, but the size of the interstices is larger and the number of them fewer. Therefore, the solubility in fcc iron is greater than in bcc. We will find that this difference is very important in heat-treating steels (Chapter 6).

In **substitutional solid solutions**, atomic size of the solute atoms and of the solvent atoms they are replacing is just as important. If the size difference is large, the solubility is correspondingly small. But other factors also are important. Hume-Rothery first defined a set of characteristics that lead to extensive solid solubility, which are known as the **Hume-Rothery rules** for substitutional solubility. According to these rules, the solute and solvent atoms must meet the following requirements:

- similar atomic size (within 15%)
- similar crystal structure
- similar valence
- similar electronegativity (the ability of an atom to attract an electron)

Sample Problem 3.1 illustrates very effectively the importance of the Hume-Rothery rules in single-phase alloys.

Figure 3.9
Annealing twins in the microstructure of 70-30 cartridge brass (500×) (From W. G. Moffatt, G. W. Pearsall, and J. Wulff, *The Structure and Properties of Materials,* Vol. 1: *Structure,* John Wiley & Sons, 1964.)

Sample Problem 3.1

Monels, **bronze**, and **brass** are alloys of copper with nickel, tin, and zinc, respectively. Using the Hume-Rothery rules as a guide, determine the comparative solubilities of Ni, Sn, and Zn in Cu. Compare your results with published room-temperature solubilities (see, for example, Figures 4.2 and 4.7 in Chapter 4).

Solution

Atomic radius	Crystal structure	Valence
Cu — 0.128 nm	fcc	$1^+, 2^+$
Ni — 0.125 nm	fcc	2^+
Sn — 0.158 nm	tetragonal	4^+
Zn — 0.133 nm	hcp	2^+

The atomic radius of Cu is 0.128 nm, so $\pm$ 15% of r_{Cu} is 0.109 nm to 0.147 nm.

Ni satisfies the Hume-Rothery rules, so the solubility should be high.

Zn satisfies atomic size variation and valence, so solubility should be good.

Sn does not satisfy any of the Hume-Rothery rules, so solubility should be limited.

Nickel indeed has complete solubility from pure Cu to pure Ni. The solubility of Zn in Cu is nearly 40%. The solubility of Sn in Cu is about 12%.

We also can add more than one alloying element to form single-phase alloys. For example, austenitic stainless steel contains iron, chromium, and nickel and Kovar alloy, used in glass-to-metal seals, contains iron, nickel, and cobalt. Even some electrical resistance heating elements are single-phase, nickel-base alloys that contain iron and chromium.

3.4   Influence of Alloying on Properties

Properties of single-phase alloys are always changed by the alloying process, but the changes are not always beneficial. For example, conductivity is always reduced by alloy additions, so only pure metals such as copper and aluminum are used for applications where only good conductivity matters. Mechanical strength is always increased by alloying, but ductility is adversely affected. Figure 3.10 shows the effect of zinc additions to copper for single-phase brass alloys. Conductivity is lowered, as expected, and strength is increased by alloying.

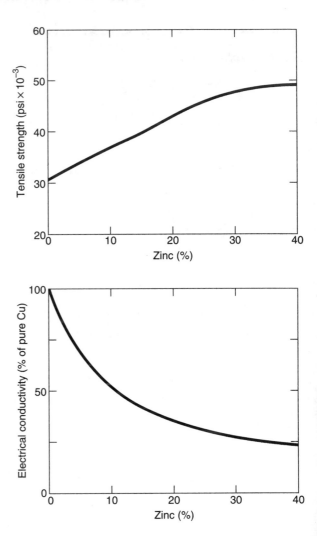

Figure 3.10
Influence of zinc additions to the ultimate tensile strength and electrical conductivity of single-phase copper alloys

One of the major contributors to mechanical strengthening by alloy additions is the difference in atomic size between solute and solvent atoms, but other contributing factors such as local atomic bonding forces also determine the degree of strengthening. We are going to learn in Chapter 9 that the solute atoms strengthen by preventing dislocation motion, a result of atomic size *difference*, not the actual atomic size. Figure 3.11 shows the effects that additions of elements have on the yield stress of copper. Most notable in this figure is tin, which is used in bronze. Bronze alloys are usually cast because they are difficult to deform, so the amount of increase in strength with small additions is important as well as how much tin is soluble.

The examples of strengthening in single-phase alloys are limited to fcc metals simply because there are no single-phase bcc substitutional alloys of any commercial interest. As a matter of fact, the effect of interstitial "alloying" is so large that we

Figure 3.11
Influence of different alloy elements on the yield strength of single-phase copper alloys

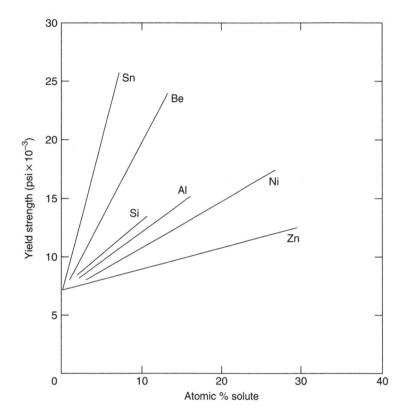

consider the **embrittlement** (loss of ductility) that results an impurity effect (see Case Study 3.1). Embrittlement is also caused by the slight tetragonal distortion of the bcc lattice, a phenomenon that we will find very important in steel heat treating.

Magnetic properties are affected by concentration of fcc single-phase, nickel-base alloys. Nickel has a low magnetic saturation, high permeability, and low coercive force. By adding iron to the nickel, the saturation can be increased with only a small increase in coercive force and small reduction in the permeability. There is another magnetic characteristic, however, that we have not looked at. It is called **magnetostriction**, which describes the interrelationship between stress and magnetism. Magnetostriction causes the atoms to repel each other, thus expanding the lattice. Loss of magnetostriction as temperature is increased counteracts normal thermal expansion and reduction in the elastic modulus. The net results are decreased thermal expansion, which is characteristic of the Fe-Ni alloy **Invar**, and isoelastic properties of similar alloys. Addition of cobalt to the Ni-Fe alloys can increase both expansion and elasticity, as shown in Figures 3.12 and 3.13. Kovar alloys that are used for matched thermal expansion required in hermetic glass-to-metal seals take advantage of the low coefficient of expansion, which matches that of borosilicate glass (Pyrex glass) up to the temperature where sealing takes place.

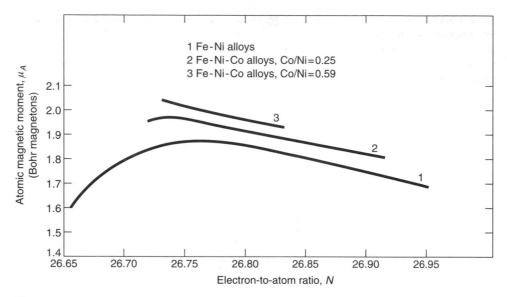

Figure 3.12
Influence of composition on magnetization of Ni-Fe and Ni-Fe-Co single-phase alloys (Electron-to-atom ratio is a convenient way to examine ternary and higher order solid solutions. It is determined from the sum of the atomic percentages of each element multiplied by its atomic number, A. $A_{Ni} = 28$, $A_{Fe} = 26$, and $A_{Co} = 27$.)

Summary

We do not have perfect metals. There are always defects in the crystal structure that may or may not affect the properties of the element. These defects can be point defects such as atomic vacancies in the crystal lattice or impurity atoms that take substitutional or interstitial positions. Impurities are usually detrimental; they decrease conductivity, decrease ductility, and decrease magnetic permeability. Dislocations are linear defects that have complicated effects on metal crystalline properties. Surface defects include grain boundaries that can strengthen metals but can reduce other properties such as magnetic permeability and conductivity. The list of pure metals that have useful applications is very small in comparison to any list of alloys. How single-phase alloys form depends on the Hume-Rothery rules for substitutional solid solutions. There are no commercially important bcc alloys, but there are a number of fcc alloys that are used. These include Cu-Ni alloys, Cu-Zn or Cu-Sn alloys, and Ni-Fe and Ni-Fe-Cr or Ni-Fe-Co alloys. Alloying always increases the strength of a metal, but magnetic property changes of the Ni-Fe alloys yield alloys with low thermal expansion or high permeability and higher magnetic saturation.

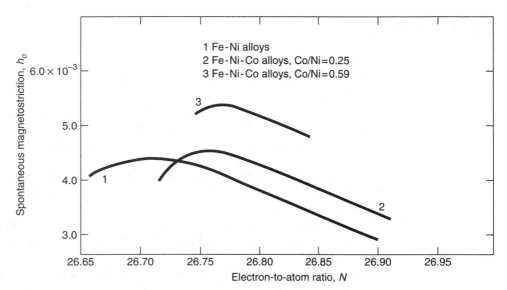

Figure 3.13
Influence of composition on magnetostriction of Ni-Fe and Ni-Fe-Co single-phase alloys (Electron-to-atom ratio is a convenient way to examine ternary and higher order solid solutions. It is determined from the sum of the atomic percentages of each element multiplied by its atomic number, A. $A_{Ni} = 28$, $A_{Fe} = 26$, and $A_{Co} = 27$.)

Terms to Remember

alloy	impurity
brass	interstitial
bronze	Invar
defects	magnetostriction
dislocation	screw dislocation
edge dislocation	solid solution
embrittlement	substitutional solid solution
grain boundary	vacancy
grain size	valence
Hume-Rothery rules	

Problems

1. Explain in your own words what point defects are.
2. Explain in your own words what dislocations are.
3. Describe what a grain boundary is.
4. Calculate the values of N for the ASTM grain sizes in Figure 3.8 using the definition in Section 3.2 and compare them to the microstructures.
5. Do grain boundaries affect magnetic saturation and magnetic coercive force? If so, how?
6. Explain the Hume-Rothery rules for substitutional solid solutions.
7. Would you expect extensive solubility of Al in Cu? Explain. (The atomic radius for Al is 0.143 nm.)
8. Hydrogen embrittlement of bcc metals occurs for a few parts per million of hydrogen. Does it matter if we think of this as ppm by weight or ppm by volume? Explain.
9. Explain why Ni-Fe-Co alloys are used for making glass-to-metal seals. (*Hint:* Glass is brittle and has low thermal expansion.)
10. Explain why strength of a metal is always increased by adding alloy elements.

4

Binary Alloys and Phase Diagrams

Single-phase alloys are the exception rather than the rule, because solubility is limited in most alloy systems. Most engineering alloys are multiphase and contain several elements. In this chapter, we will examine the behavior of **binary alloys** (alloys containing two elements) that are at **equilibrium**, that is, they are stable. Once the solubility of one element in another is exceeded, a second phase is formed, and both phases coexist with different compositions and structures. Of course, the properties of these two phases are different and the properties of the alloy itself will depend on the composition and the amount of each phase in addition to other factors such as size and distribution of the phases.

In order to understand binary alloys, we must first look at the most stable, or equilibrium, condition. Equilibrium is a thermodynamic concept, but one that we can accept and do not need to examine in any detail. All we must remember is that the lowest energy state is the most stable. Much of our understanding of alloys will depend on interpretation of *equilibrium phase diagrams*, which are graphical representations of the phases present in a binary alloy at various temperatures, compositions, and pressures. We will use these diagrams to select, understand, and predict alloys and their behavior.

Some of the important information that can be obtained from phase diagrams are:

- melting conditions of alloys
- number, amount, and composition of phases present at a particular temperature
- solubility of one element in another

4.1   *The Gibbs Phase Rule*

A pure substance can exist in solid, liquid, or vapor phases, depending on the temperature and pressure. We know that some solids, such as iron, can also have different crystal structures at different pressures and temperatures. The number of phases in equilibrium in any system can be determined by an equation known as the **Gibbs phase rule**, named after the nineteenth century physicist who first derived it from thermodynamic considerations. The Gibbs phase rule is

$$P + F = C + 2$$

where P is number of phases that can coexist at one time in a given system, C is the number of components or elements in the alloy system, and F is the degrees of freedom. The degrees of freedom represent the number of variables (pressure, temperature, or composition) that can be changed without changing the number of equilibrium phases in the system.

We can better understand the significance of the Gibbs phase rule and begin to understand phase diagram construction by looking at the difference in **cooling curves** for a pure metal and a binary alloy. Figure 4.1 depicts these cooling curves during solidification. For the pure metal, $C = 1$ and if we fix the pressure at 1 atmosphere (which is standard procedure), the Gibbs phase rule states that $F = C + 1 - P$. When $P = 2$ (both solid and liquid are present), then $F = 0$. We cannot change the temperature until the liquid phase solidifies; thus, there is a **thermal arrest** at the single melting point for the pure metal in Figure 4.1a.

In Figure 4.1b, however, $C = 2$, and the Gibbs phase rule states that $F = 1$ when both liquid and solid phases are present and temperature can be changed without any thermal arrest. The inflections in the cooling curve that represent the beginning and completion of solidification are termed, respectively, the **liquidus** and **solidus** temperatures for the alloy. The only phase present at temperatures above the liquidus is liquid and the only phase present below the solidus is solid.

Cooling curves are used to construct phase diagrams, changing the composition of the alloy for each test. Let's examine the **copper-nickel phase diagram** in Figure 4.2 for example. In Chapter 3, we learned that copper and nickel are the best match to satisfy the Hume-Rothery rules for solid solubility, and these elements

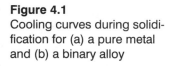

Figure 4.1
Cooling curves during solidification for (a) a pure metal and (b) a binary alloy

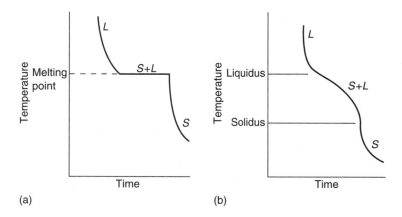

indeed form a complete series of single-phase solid solutions from pure copper to pure nickel. The Cu-Ni phase diagram in Figure 4.2 shows that only pure Cu and pure Ni have single melting points and all alloys have a two-phase region between the liquidus and solidus.[1]

We will return to the Gibbs phase rule for more complicated phase diagrams, but let's look at the information we can obtain from any phase diagram, using the composition and a **tie line**, a constant temperature line connecting two single phases, as identified in Figure 4.3 (for a hypothetical A-B alloy). In this figure, if we have an alloy of composition w_0 at temperature T, the composition of the liquid phase is given by that corresponding to the intersection between the tie line and the liquidus, or w_L, and the composition of the solid phase is given by that corresponding to the intersection between the tie line and the solidus, or w_S. We can also determine the amounts of each phase present by a simple material balance. However, it is easier to refer to the inverse lever rule, which is the consequence of a material balance. The **inverse lever rule** simply states that

$$F_S = \frac{W_o - W_L}{W_S - W_L}$$

where F_S is the weight fraction of the solid phase. The weight fraction of the liquid phase is simply given by

$$F_L = 1 - F_S$$

$$\text{or } F_L = \frac{W_S - W_o}{W_S - W_L}$$

This rule will be very helpful in understanding property-composition relationships and in beginning to understand nonequilibrium behavior.

---

[1] It is traditional to use Greek letters to identify single phases. Common letters used are alpha (α), beta (ß), gamma (γ), delta (δ), epsilon (ε), kappa (κ), phi (ϕ), and theta (θ).

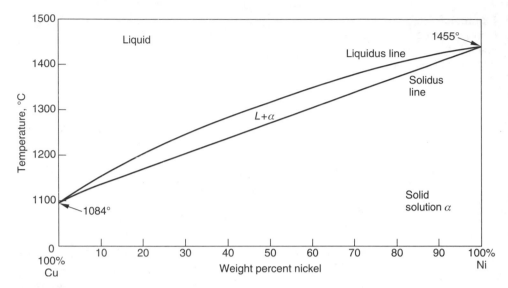

Figure 4.2
The copper-nickel phase diagram
(From *Metals Handbook,* Vol. 8: *Metallography, Structures and Phase Diagrams,* 8th ed.,
ASM International, 1986.)

Figure 4.3
Hypothetical A-B alloy
demonstrates tie lines and
inverse lever rule

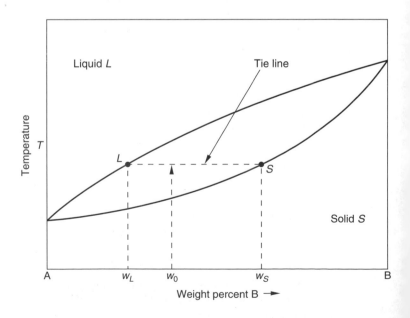

Sample Problem 4.1

An alloy of copper and nickel contains 23 weight percent Ni. At 1200°C, what are the compositions of the liquid and solid phases and what is the weight percent of solid and liquid?

Solution

Using Figure 4.2, the tie line at 1200°C shows the liquid composition is 19% Ni and 81% Cu and the solid composition is 30% Ni and 70% Cu.

$$F_S = \frac{W_o - W_L}{W_S - W_L} = \frac{23 - 19}{30 - 19} = .364, \text{ or } 36.4\%$$

$$F_L = 1 - F_S = 1 - .364 = .636, \text{ or } 63.6\%$$

4.2 Invariant Reactions

Most binary alloy systems have limited solid solubility, that is, a second phase appears if we exceed the solubility limit of one element in the other. Solubility is affected by temperature, but when melting begins, there will be three phases present. Applying the Gibbs phase rule,

$$F = C + 1 - P$$
$$= 2 + 1 - 3$$
$$= 0$$

We call a reaction that has no degrees of freedom an **invariant reaction**. Figure 4.4 illustrates the cooling curves for the two invariant reactions that involve melting or, conversely, solidification. In the **eutectic** reaction, a single liquid phase forms two solid phases upon solidification; there is a thermal arrest as the three phases coexist.

Figure 4.4
Cooling curves during solidification of binary alloys displaying invariant reactions: (a) eutectic reaction, (b) peritectic reaction

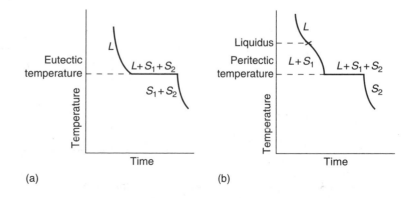

In the **peritectic** reaction, a single liquid phase combines with a single solid phase to form a second solid phase with different composition upon cooling. Again, a thermal arrest occurs while the three phases coexist.

In the case of the pure metal, a thermal arrest signified a single melting point on the phase diagram. For eutectic and peritectic reactions, however, there are different compositions of the solid and liquid phases. On the phase diagram, we must have a single melting temperature extending between the limits of solubility for the respective phases, as illustrated in Figure 4.5. Solubility limits always increase with temperature, so the lines separating single- and two-phase regions, called **solvus** lines, always curve toward the two-phase region as temperature is increased.

By convention, we describe these invariant reactions upon cooling. Thus, the eutectic reaction is described as

$$L(C_E) \rightarrow S_1(C_{S1}) + S_2(C_{S2}) \text{ at } T_E$$

Figure 4.5
(a) Eutectic invariant reaction on a phase diagram and
(b) peritectic invariant reaction on a phase diagram

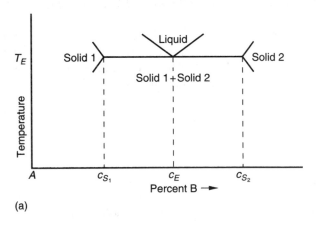

(a)

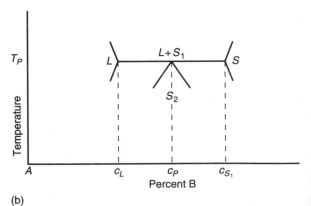

(b)

and the peritectic reaction is described as

$$L(C_L) + S_1(C_{S1}) \rightarrow S_2(C_{S2}) \text{ at } T_P$$

where concentrations, C, are given in percent of the binary alloy addition.

Eutectic alloys find applications because of the lower melting temperature of the alloy as compared to the pure metals. Perhaps the most common example is the **lead-tin phase diagram**, reproduced in Figure 4.6. Lead-tin alloys are used for joining or soldering metals, as we will learn in Chapter 10.

Sample Problem 4.2

Refer to Figure 4.6 as you perform the following operations.

 a. Write the eutectic reaction.

 b. Draw cooling curves for liquid alloys of the compositions 10% Sn, 40% Sn, 61.9% Sn, and 80% Sn.

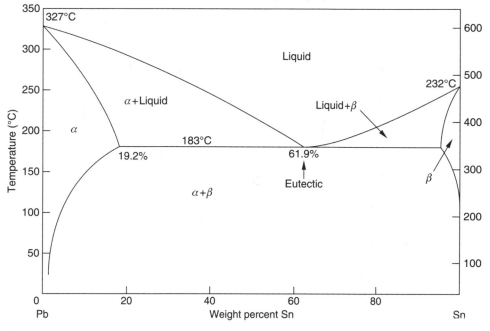

Figure 4.6
The lead-tin phase diagram
(From *Metals Handbook,* Vol. 8: *Metallography, Structures and Phase Diagrams,* 8th ed., ASM International, 1986.)

Solution

a. $L(61.9\%\ Sn) \rightarrow \alpha(19.2\%\ Sn) + ß(97.5\%\ Sn)$

b.

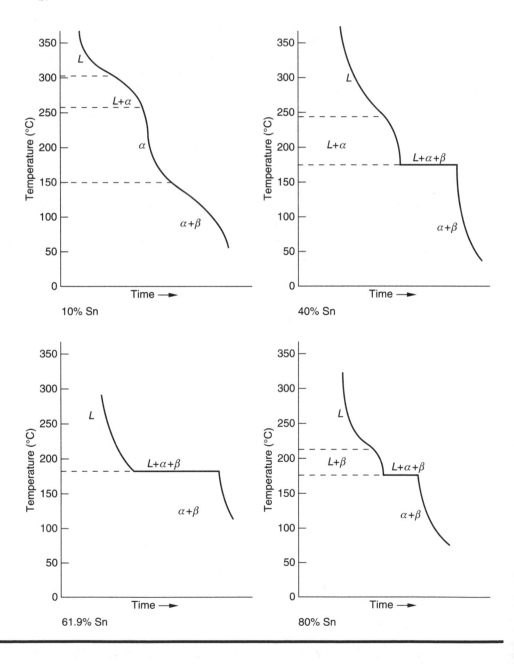

There are two other invariant reactions that we commonly encounter: the eutectoid and peritectoid reactions. These appear to be the same as the eutectic and peritectic reactions, with one very important difference — only solid phases are involved. Thus the **eutectoid** reaction can be described as

$$S_1(C_E) \rightarrow S_2(C_{S2}) + S_3(C_{S3}) \text{ at } T_E$$

and the **peritectoid** reaction is described as

$$S_1(C_1) + S_2(C_{S2}) \rightarrow S_3(C_{S3}) \text{ at } T_P$$

where concentrations, C, are given in percent of the binary alloy addition.

4.3   Phase Diagrams of Common Commercial Alloys

Copper-nickel alloys that have found commercial use include **cupronickels**, which are copper-base with 10% and 30% Ni additions, and **monels**, which are nickel-base with 30% Cu additions. The wide variation in composition of these alloys is due to the complete solid solubility range. We now know enough to examine more complicated phase diagrams.

Figure 4.7 shows the **copper-zinc phase diagram**. The solubility of zinc in copper is limited to about 40% by the Hume-Rothery rules, which we studied in Chapter 3. For higher zinc compositions, the phase diagram appears to be quite cluttered. (Note that the dashed lines in this figure indicate uncertainty caused by the very long times necessary to reach equilibrium at the lower temperatures, which is a common problem in phase diagram determination.)

There are a number of brass alloys containing copper and zinc that are commercially important. These include red brass (15% Zn), cartridge brass (30% Zn), yellow brass (35% Zn), and **Muntz metal** (40% Zn). Note that all of these are single-phase α alloys. It is important that we recognize that commercial alloys must be reproducible in composition and properties and must be able to be shaped into a product. It is this reproducibility and manufacturability that limits many commercial alloys to single-phase materials. One very important exception to this is the **iron–iron carbide phase diagram**, which is the basis for the iron and steel industry.

Sample Problem 4.3

Identify all the invariant reactions in the Cu-Zn phase diagram (Figure 4.7).

Figure 4.7
The copper-zinc phase diagram
(From *Metals Handbook,* Vol. 8:
*Metallography, Structures and
Phase Diagrams,* 8th ed., ASM
International, 1986.)

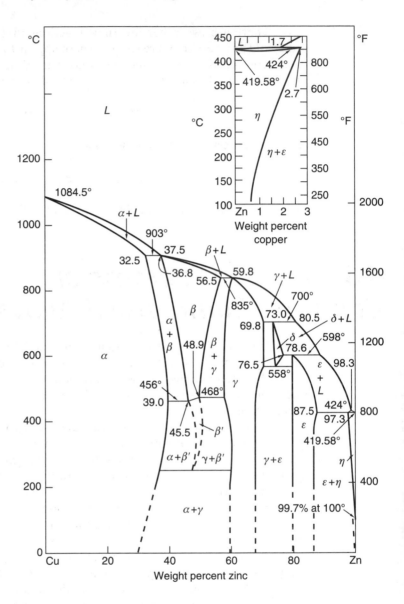

Solution

1. $T = 903°C$; peritectic, $\alpha(32.5\%\ Zn) + L(37.5\%\ Zn) \rightarrow ß(36.8\%\ Zn)$
2. $T = 835°C$; peritectic, $ß(56.5\%\ Zn) + L(59.8\%\ Zn) \rightarrow \gamma(59.7\%\ Zn)$
3. $T = 700°C$; peritectic, $\gamma(69.8\%\ Zn) + L(80.5\%\ Zn) \rightarrow \delta(73.0\%\ Zn)$
4. $T = 598°C$; peritectic, $\delta(76.5\%\ Zn) + L(87.8\%\ Zn) \rightarrow \varepsilon(78.6\%\ Zn)$
5. $T = 558°C$; eutectoid, $\delta(73.9\%\ Zn) \rightarrow \gamma(70.3\%\ Zn) + \varepsilon(78.6\%\ Zn)$
6. $T = 419.6°C$; peritectic, $\varepsilon(87.5\%\ Zn) + L(98.6\%\ Zn) \rightarrow \eta(97.3\%\ Zn)$
7. $T = 245°C$; eutectoid, $ß'(46.5\%\ Zn) \rightarrow \alpha(36.0\%\ Zn) + \gamma(59.5\%\ Zn)$

4.4   The Iron–Iron Carbide (Fe-Fe₃C) Phase Diagram

Carbon is a nonmetallic element that has limited solubility in iron despite its small atomic size. In addition, iron and carbon form a covalently bonded compound, Fe_3C, that we call iron carbide, or **cementite**. This compound contains 6.7% carbon by weight. It is the Fe-Fe₃C phase diagram, shown in Figure 4.8, that we will find to be critical in understanding the composition, heat treatment, manufacturability, and other aspects of ferrous alloys. The term **ferrous** denotes that iron is the major constituent and ferrous alloys include steels and cast irons. We will learn much more about them in Chapter 7, but here we will only be concerned about the phase diagram itself.

   Steels contain up to about 1% carbon by weight. Figure 4.8 shows that these alloys have high melting ranges and a high-temperature peritectic reaction, which is necessary according to the Gibbs phase rule because of the high-temperature polymorphic transformation in iron. The high-temperature bcc crystal structure is called the delta (δ) phase; the low-temperature bcc crystal structure (stable at room tem-

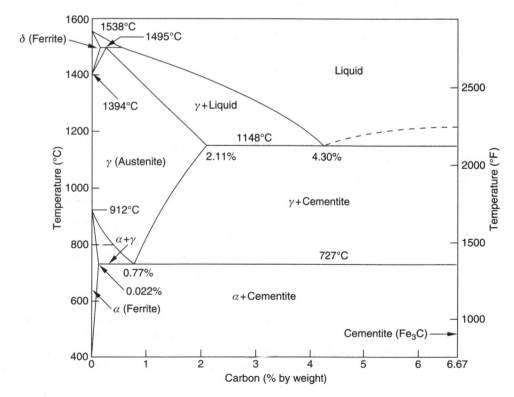

Figure 4.8
The iron–iron carbide phase diagram

perature) is called the alpha (α) phase. Both structures are referred to as ferrite, but convention dictates the use of the term **ferrite** for the α phase and δ-ferrite for the δ phase. The phase in between has a stable fcc crystal structure and is called **austenite**.

The transformation from austenite to ferrite on cooling is a eutectoid reaction, with the eutectoid composition 0.77% carbon and the eutectoid temperature 727°C. The importance of this transformation cannot be overemphasized, as we will learn shortly. The other important ferrous alloys are cast irons, also described in more detail in Chapter 7. **Cast irons** take advantage of the low melting temperature related to the eutectic reaction at 4.3% carbon ($T_E = 1148$°C) because they can be cast to shape very economically.

There are a number of features of the Fe-Fe$_3$C phase diagram that are important. Note that the solubility of carbon is much higher in austenite than in ferrite. This occurs because the small carbon atom forms an interstitial solid solution. Although the bcc structure is less dense than the fcc structure, remember from Chapter 2 that the size of the interstices is larger in the fcc structure. The larger solubility is advantageous in heat treating because we can heat steel into the austenitic range and have a uniform single phase as our starting point, then control the transformation upon cooling. At room temperature, all steels have more than one phase because of the limited solubility of carbon in ferrite. The size and distribution of this second phase affect the properties of the steel, particularly the strength for which steel is noted. We refer to such phenomena as size and distribution of a second phase or the grain size of the principal phase as microstructure. Microstructure is observable with a microscope for flat polished sections of metal that are etched to reveal the different features.

4.5   Microstructure and Invariant Reactions

Microstructure includes other features of etched metal surfaces besides the grain size and size and distribution of second phases. For example, annealing twins within the grains in brass were demonstrated in Figure 3.9. The microstructure that we see for room-temperature alloys is formed by the lowest invariant reaction temperature. For example, the lead-tin microstructure of a 60% Sn-40% Pb solder is formed by the eutectic reaction at 183°C, with only the composition of the phases being altered by cooling to room temperature. Similarly, the microstructure of steel is controlled by the eutectoid reaction at 727°C, with little change even in composition upon cooling to room temperature. (The microstructure of single-phase alloys, on the other hand, is not changed once solidification takes place, unless we mechanically deform it, as we will learn in Chapter 9.)

If we examine what happens on a microscopic scale when the single phase transforms on cooling to two phases of widely different compositions, we see that a great many atoms must move over very large distances before equilibrium is reached. It turns out to be much easier for many atoms to move short distances; they form a platelike structure that is very stable, despite the fact that equilibrium is not achieved. We refer to this microstructure as eutectic microstructure or eutectoid microstructure and it has the composition and amounts determined by the tie line defined by the invariant reaction. In the case of steel, the eutectoid microstructure is named **pearlite**.

Sample Problem 4.4

  a. Using Figure 4.8, calculate the percentage of ferrite and carbide just below the eutectoid temperature in an iron-carbon alloy containing 0.3% carbon.

  b. Calculate the percentage of ferrite and carbide just below the eutectoid temperature in an iron-carbon alloy having the eutectoid composition (this is pearlite).

  c. Recalculate part *a* in terms of ferrite and pearlite instead of ferrite and carbide.

Solution

Using the inverse lever rule,

a.
$$\%\alpha = \frac{6.67 - 0.30}{6.67 - 0.02} \times 100$$
$$= 95.8\%$$
and $\%Fe_3C = 4.2\%$

b.
$$\%\alpha = \frac{6.67 - 0.77}{6.67 - 0.02} \times 100$$
$$= 88.7\%$$
and $\%Fe_3C = 11.3\%$

c.
$$\%\alpha = \frac{0.77 - 0.30}{0.77 - 0.02} \times 100$$
$$= 62.7\%$$
and $\%$ pearlite $= 37.3\%$

In Sample Problem 4.4, the amounts of ferrite and carbide are the same but the distribution and, therefore, the microstructures are different. Because it takes an infinitely long time for atoms to move through the lattice to reach equilibrium, we never see a fine distribution of carbide in ferrite. We do, however, encounter the ferrite and pearlite microstructures frequently. In these cases, we distinguish the ferrite that is free from that tied up with the carbide as pearlite. Free ferrite is also called *proeutectoid* ferrite.

4.6 Polynary Alloys and Phase Diagrams

Most commercial alloys are **polynary alloys** — they contain more than two components. Although we do occasionally encounter ternary phase diagrams, they are much more complicated than binary phase diagrams and we usually find it much easier to qualitatively describe the effects of additional alloying elements on the binary phases. For example, many steels have additions of silicon, manganese, nickel, molybdenum, chromium, and other alloys added in small amounts. We can examine the effects of these additions on how they shift the eutectoid reaction in the binary Fe-Fe₃C phase diagram. Figure 4.9 shows the effect of alloy additions on the eutectoid temperature. Only manganese and nickel lower the eutectoid temperature; other alloy additions raise it. We refer to manganese and nickel as austenite **stabilizers** and other elements as ferrite stabilizers because of the shift. Nickel is most important in this regard because it can stabilize the fcc austenite to very low temperatures. Thus we have important commercial iron-base alloys such as austenitic stainless steels and Invar-type specialty alloys whose properties are derived from iron in the fcc lattice. (Invar, which is Fe-36% Ni, has near zero expansion coefficient from room temperature to its Curie temperature.)

Figure 4.9
The effect of alloying elements on the eutectoid temperature of Fe-Fe₃C alloys

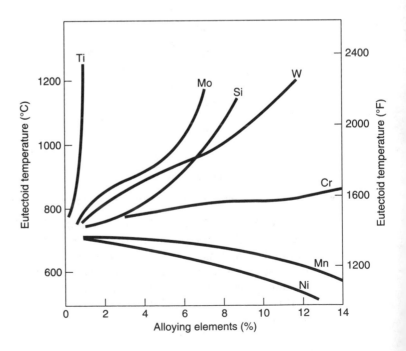

Summary

Metal alloys typically are multiphase alloys, containing more than one single phase. How the phases coexist has been described using the Gibbs phase rule, which tells whether the number of phases must be altered or not in order to change the temperature or pressure of a system. Cooling curves are affected by and phase diagram construction must conform to this rule. Phase diagrams are important for alloy selection and understanding property dependence on composition. Invariant reactions, which require thermal arrest upon transformation, include eutectic reactions that lower the melting point of alloys in comparison to pure elements, the similar eutectoid reactions that involve only solid phases, the peritectic reactions where liquid combines with one solid phase upon cooling to form a second solid phase of different composition, and the similar peritectoid reactions that involve only solid phases. Phase equilibria and microstructure of commercial alloys are dependent on the amount and composition of the phases, which is described by the inverse lever rule. Effects of polynary alloy additions on phase equilibria of commercial alloys were also shown.

Terms to Remember

austenite

binary alloy

brass

cast iron

cementite

cooling curves

copper-nickel phase diagram

copper-zinc phase diagram

cupronickel

equilibrium

eutectic

eutectoid

ferrite

ferrous

Gibbs phase rule

Invar

invariant reaction

inverse lever rule

iron–iron carbide phase diagram

lead-tin phase diagram

liquidus

microstructure

monel

Muntz metal

pearlite

peritectic

peritectoid

polynary alloys

solidus

solvus

stabilizer

steel

thermal arrest

tie lines

Problems

1. Given a monel alloy containing 65% Ni and 35% Cu,
    a. what is the temperature of the liquidus? of the solidus?
    b. what are the compositions and amounts of the liquid and of the solid at 1350°C?

2. In the Pb-Sn phase diagram (Figure 4.6), what are the maximum solubilities of Sn in Pb and Pb in Sn?

3. An alloy has the composition 30% Sn, 70% Pb.
    a. Draw its cooling curve from 350°C to room temperature.
    b. What are the compositions and amounts of the phases at 200°C?

4. A Pb-Sn alloy contains 30% *L* and 70% *α* at 200°C. What is its composition in weight percent?

5. Identify the invariant reactions in the accompanying Al-Ni phase diagram.

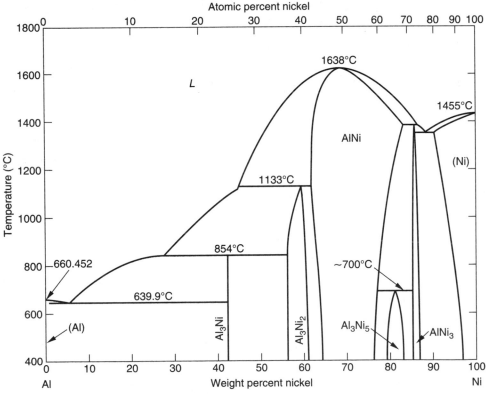

(From *Metals Handbook*, Vol. 8: *Metallography, Structures and Phase Diagrams*, 8th ed., ASM International, 1986.)

6. Plot the cooling curve for an alloy of copper and zinc containing 80% Zn. Why is this alloy not promising as a commercial alloy?

7. What phases are present in Cu-Zn alloys at 500°C? Give the solubility limits of single phases.

8. What are some practical implications of eutectic alloys?
9. Plot the weight fraction of phases versus temperature for a plain carbon steel containing 0.40% C.
10. Define the phases present in a plain carbon steel containing 0.40% C. With your knowledge gained from Chapter 2, plot the volume change in this alloy as it cools from the liquid phase.
11. On graph paper, construct the equilibrium A-B phase diagram from the following information:

    Pure A melts at 700°C.

    Pure B melts at 900°C.

    The solubilities at 100°C are B in A, 3%; A in B, 3%.

    Invariant Reactions

    At 750°C, peritectic, $L(70\% \text{ B}) + \gamma(95\% \text{ B}) \rightarrow \text{ß}(85\% \text{ B})$.
    At 500°C, eutectic, $L(40\% \text{ B}) \rightarrow \alpha(10\% \text{ B}) + \text{ß}(60\% \text{ B})$.
    At 300°C, eutectoid, $\text{ß}(80\% \text{ B}) \rightarrow \alpha(5\% \text{ B}) + \gamma(95\%\text{B})$.

12. Label the phase fields and identify the invariant reactions for the accompanying equilibrium A-B phase diagram.

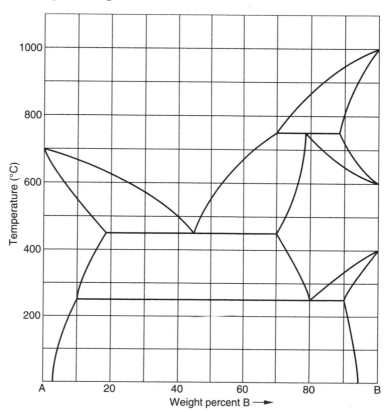

5

Melting and Solidification

We have learned about atomic structure, bonding, solid solutions, multiple phases, grain boundaries, and microstructure in order to relate the properties to the characteristics of metal alloys. Extension of what we know can move in several different directions in order to complete our understanding of material properties. The approach in this and subsequent chapters will be to look at the ways in which we can shape metal alloys and control their microstructure and properties. Although we can cast them to shape in a mold, deform them to shape, or shape them by removing metal, we will look only at the first two methods. Machining or material removal does not alter microstructure or properties to any extent, and machining processes themselves are the subject of numerous textbooks.

5.1 Production of Metal Alloys

Almost all metallic alloys that are commercially important are made by mixing the elements as liquids because they are *miscible*, that is, they form a single liquid phase whose composition is therefore uniform. In some cases, particularly for many steels, the ores are refined, producing molten metals that are immediately alloyed — an economic efficiency. In other cases, purity of the alloys is critical, so we must purify the elements first and then alloy them by heating, as with Ni-Fe magnetic alloys. Although we will not be studying alloy production in detail, some exposure to the processes is important for our overall understanding of the relationships between materials and their properties.

Steelmaking begins with iron ore that is reduced by carbon and refined by limestone slag in a **blast furnace**, such as the one shown schematically in Figure 5.1. This process of reduction and refinement in the blast furnace is continuous and results in molten pig iron, which contains high amounts of carbon and silicon. Most pig iron is transported from the blast furnace in the molten state to open hearth or **basic oxygen furnaces** where it is further refined into steel. The basic oxygen process, shown schematically in Figure 5.2, removes the carbon and silicon to the desired levels of the steel being produced. As much as 250 tons of steel can be refined in less than an hour. Most steels made by the basic oxygen process are used for plate, sheet, or structural beams of many shapes.

Electric furnaces are used in steelmaking either when products are made in small quantities or when process control is critical. Chemical refinement takes place,

Figure 5.1
Schematic of a blast furnace

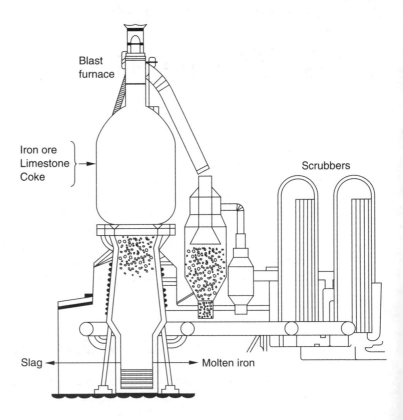

Blast
furnace

Iron ore
Limestone
Coke

Scrubbers

Slag ◄—   —► Molten iron

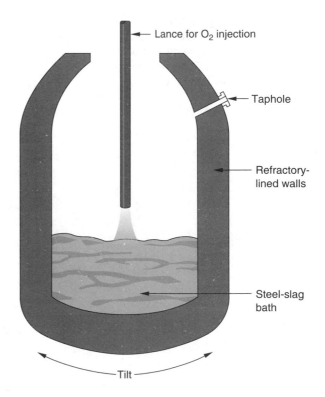

Figure 5.2
Cross section of a basic
oxygen furnace

Lance for O$_2$ injection

Taphole

Refractory-
lined walls

Steel-slag
bath

Tilt

but the beginning material is usually a solid (such as scrap) that is heated to the liquid phase. Large quantities of steel are made by three-phase **electric arc furnaces**, such as the direct arc furnace illustrated in Figure 5.3a. Figure 5.3b shows a 25-ton furnace in the process of tapping the batch heat. High-quality products, which are used for stainless and other specialty steels, are achieved in these furnaces. Another type of electric furnace is the **induction furnace**, shown in Figure 5.4. In this furnace, solid metal is added to the **crucible**, or liner, and high-frequency electric current is passed through the water-cooled copper solenoid coils. This current induces a magnetic field that in turn induces electrical resistance heating of the metal. Little refining can be done by this method. Induction melting is used for such metals as copper and specialty alloys and for experimental quantities.

Metals that do not react with crucible material and that have sufficiently low melting points can be melted in **fuel-fired furnaces** such as that depicted in Figure 5.5. These furnaces are suitable for small heat sizes, and crucibles can be lifted from the furnace and then poured directly into molds. Solders, aluminum, magnesium alloys, and some copper alloys are melted in fuel-fired furnaces.

Some metals and alloys cannot be melted by any of these techniques because impurities cannot be adequately controlled. **Vacuum melting** is necessary for such alloys, and there are two types that are used extensively. For better control of inclusions that end up in the solid, vacuum induction melting is used. For other materials, however, the properties are so sensitive to oxygen that alloying must be done by

Figure 5.3
Electric arc furnace:
(a) schematic and
(b) 25-ton furnace
being tapped into
ladle

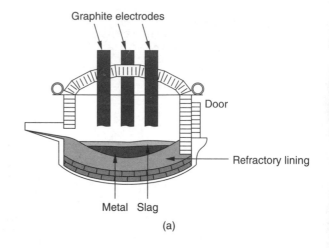

Graphite electrodes

Door

Refractory lining

Metal   Slag

(a)

(b)

Figure 5.4
Schematic of an induction
furnace

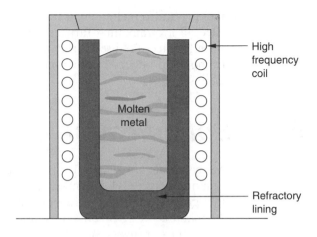

High
frequency
coil

Molten
metal

Refractory
lining

Figure 5.5
Schematic of a fuel-fired
crucible furnace

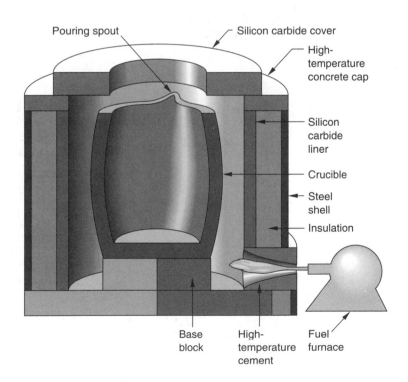

Pouring spout

Silicon carbide cover

High-
temperature
concrete cap

Silicon
carbide
liner

Crucible

Steel
shell

Insulation

Base
block

High-
temperature
cement

Fuel
furnace

melting and remelting with no contact with any crucible. Figure 5.6 illustrates the vacuum arc remelting furnace, which is a vacuum-consumable electrode furnace. Such furnaces are used for titanium and exotic alloys such as Nb-Ti or Zr alloys that cannot be produced by other means.

5.2   *Solidification of Alloys*

Many metals are shaped by casting molten alloys into molds, the process that the foundry industry is based upon. The number of foundries has shrunk in recent years, though, and the study of foundry practices is not within the scope of this textbook. As manufacturing practitioners, however, we should be aware of the properties of cast metals or ingots. **Ingots** are made in permanent molds that are later deformed, but the microstructure, chemistry distribution, and porosity that we encounter in ingots is similar to those of shaped castings.

    Solidification of metals is accompanied by a large decrease in volume as atoms loosely bonded in the liquid phase become aligned in the long-range ordered crys-

Figure 5.6
Schematic of a vacuum
arc remelting furnace

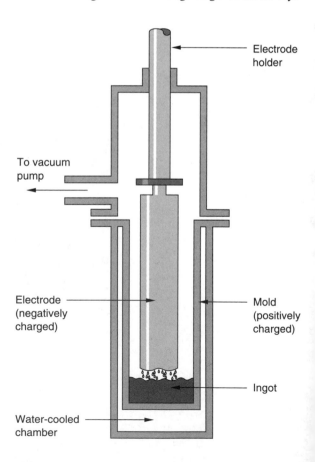

Electrode
holder

To vacuum
pump

Electrode
(negatively
charged)

Mold
(positively
charged)

Ingot

Water-cooled
chamber

talline lattice. We describe this solidification as a **nucleation** and **growth** of the solid phase(s), in which the long-range order of the solid forms first in a nucleus that then grows in the direction opposite to that in which the heat is removed. Such a process results in characteristics that have a profound influence on the properties of the cast solid. These characteristics — cast microstructure, chemical segregation, and porosity or **microporosity** — cause nonuniform properties and stress concentration that make castings inferior to deformed metals.

5.2.1   Cast Microstructure

The cast microstructure results from the nucleation process. Nuclei of the solid phase form where the energy will be reduced the most, that is, where the temperature is the coldest — at the edges in contact with the mold. We call the area where nuclei form the **chill zone**. Many grains form but do not grow because heat is removed quickly. When the heat removal slows because it becomes limited by the thickness and thermal conductivity of the solid, then grains begin to grow, forming a columnar shape that grows in the direction opposite to that in which heat flows. When the thickness becomes large enough, then there is the opportunity for additional nucleation rather than growth of the existing grains, and the grain structure again changes. We call these internal grains **equiaxed grains** because they are nearly spherical in shape. Figure 5.7 shows the microstructure of a cross section of the center of a square ingot that has solidified in this manner.

Figure 5.7
Schematic of cast microstructure of a square ingot

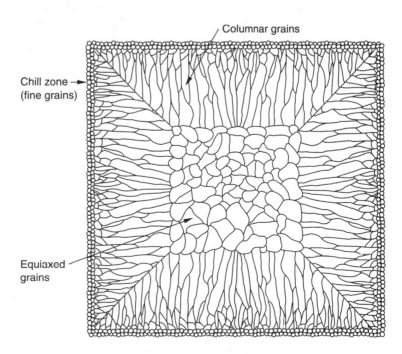

Columnar grains

Chill zone →
(fine grains)

Equiaxed
grains

The strength and ductility of cast metals decrease as we go from the surface inward, but the decrease is much greater for ductility. In addition, the properties of **columnar grains** are better parallel to the growth direction than transverse to it. These changes in properties are not only attributed to the microstructure, however; there are chemical changes and microporosity changes that are also associated with solidification. For example, high melting point nonmetallic impurities that float in the liquid metal are repelled when solidification takes place. These impurities end up as **nonmetallic inclusions** in the solid when all solidification is completed, with higher concentration in the center than at the outside. Figure 5.8 shows the appearance of nonmetallic oxide inclusions in columnar and equiaxed zones of air-melted and vacuum-melted steels. The properties are higher for the vacuum-melted steels, of course, because of the reduced impurities in the microstructure.

5.2.2   *Chemical Segregation*

In the study of alloy phase diagrams, we looked only at equilibrium conditions, where an infinite amount of time permitted the movement of atoms to form the different phases. In the real world, we do not have such opportunity and end up with chemistry differences that we call **segregation**, or **coring**. For example, Figure 5.9 shows the distribution of alloy additions to steel in an equiaxed grain near the center of the ingot.

Figure 5.8
Nonmetallic oxide inclusions:
(a) columnar region, air melt;
(b) equiaxed region, air melt;
(c) columnar region, vacuum melt; and (d) equiaxed region, vacuum melt

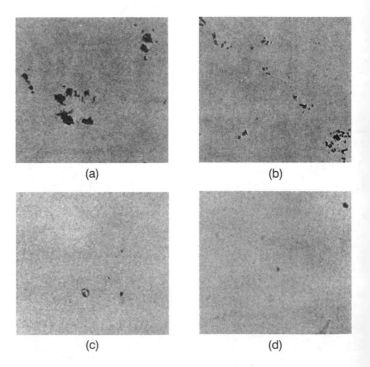

(a)   (b)

(c)   (d)

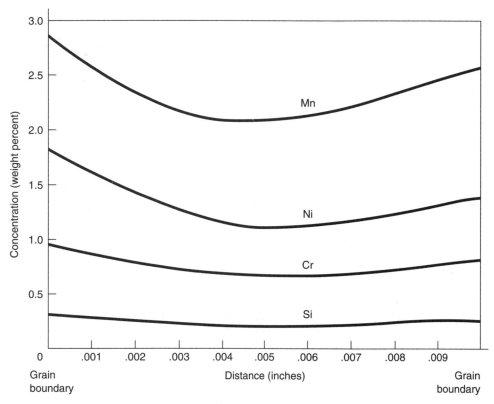

Figure 5.9
Change in composition of alloying elements in an equiaxed grain of an as-cast steel

Segregation results from the redistribution of solutes during the solidification process because there is insufficient time for atoms to move into equilibrium positions. We can understand this more easily by considering the tie line between the solidus and liquidus of a binary alloy in Figure 5.10a. The ratio of the concentration of the solid to the concentration of the liquid is known as the **distribution coefficient**, k. In an alloy of composition, c_o, the composition of the first solid formed is c_S and the liquid is enriched to c_L, as shown schematically in Figure 5.10b. When factors such as density differences, mixing in the liquid state, sediment settling, and so on are considered, the concentration profile is similar to that shown in Figure 5.9, where solidification started at the center of the equiaxed grains. The amount of difference from the center to the grain boundary is a function then of the size of k. For example, k is higher for manganese and nickel in iron than for silicon and chromium in iron.

The effect of segregation on properties is the same as that of impurity effects on properties. Segregation is one of the factors that make cast metals inferior.

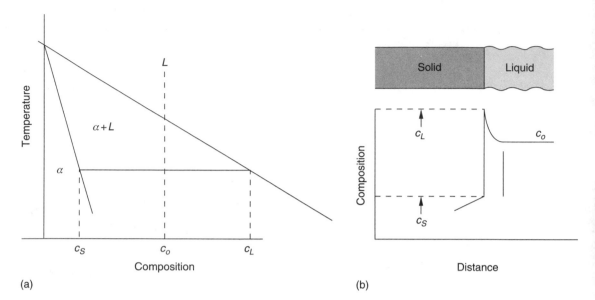

Figure 5.10
Nonequilibrium distribution of an alloy element during solidification of a binary alloy:
(a) phase diagram, (b) composition of a solidifying liquid

5.2.3   *Porosity Formed in Solidification*

When the liquid phase solidifies, the solid phase occupies less volume, resulting in **shrinkage**. Remaining liquid must travel ever-larger distances as solidification proceeds, but the net effect is a solid that occupies less space than did the liquid. Gross shrinkage, or **macroporosity**, can be compensated by altering the shape of the mold, for example, by adding larger sections called **risers** that take longer to solidify and thereby continue to feed the mold. Risers are cut away after solidification, leaving solid castings that have few voids. We see the effect of riser control of gross shrinkage during casting in the sketch of Figure 5.11. On a microscale, however, it is impossible to prevent some microporosity from occurring, particularly in grain boundaries where it acts as stress-concentrating microcracks.

Microporosity can also be caused by the release of gases absorbed during the melting processes. The liquid phase can dissolve much more gas than the solid phase can. Wherever we encounter large amounts of absorbed gas, then, we have to be sure it is released before solidification begins. Aluminum alloys are particularly susceptible to hydrogen absorption, and despite methods practiced for its removal immediately before casting, aluminum castings are rarely used for tensile applications because of microporosity.

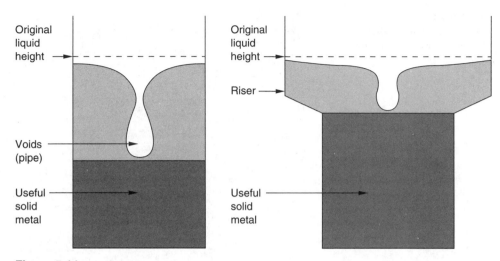

Figure 5.11
(a) Solidification shrinkage and (b) correction by adding a riser that is designed to be the last section to solidify

Case Study 5.1

Failure of a Cast Steel Bolt

SW Castings produces aluminum alloy die castings. In the melting area, alloys are prepared in a fuel-fired crucible furnace (see Figure 5.5), then the crucibles are lifted by means of a small crane suspended from a ¾-in. diameter cast steel bolt (¾-10 UNC threads). Safe load for a steel bolt of this size is slightly more than 3000 lb, much higher than the furnace capacity of 1000 lb. Nevertheless, some five years after installing the system, the bolt failed catastrophically, dropping the crucible full of molten metal. Fortunately, no one was hurt, but production was interrupted for some time.

The section of bolt that remained in the ceiling, shown in Figure 5.12, was brittle. Scanning electron microscopy revealed numerous microcracks (Figure 5.13), but no evidence of any fatigue, which was initially considered possible. Chemical analysis revealed the steel to be a cast A148 steel. A bolt of the same origin was machined and tested, with the following results:

Sample	0.2% Y.S. (psi)	UTS (psi)	%El	%RA	Impact (ft-lb)
A148 standard	60,000	90,000	20*	40*	40
Bolt	58,800	100,000	18.1	30.0	2.5

These results show that the cast steel is inferior and should not have been used for the application. Replacement with a forged steel bolt has been successful.

Figure 5.12
Brittle fracture surface of
the bolt section remaining
in the ceiling

Figure 5.13
SEM micrograph of the
fracture surface of the bolt
(160×). Arrows indicate
microcracks.

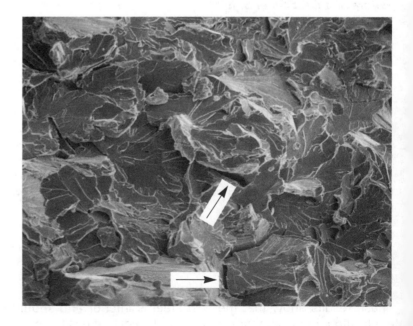

5.3 Microstructure Resulting from Solidification of Eutectic Alloys

The microstructure that develops during solidification of eutectic alloys is unique. The single-phase liquid forms two phases, a process that requires extensive redistribution of the atoms (diffusion is covered in Chapter 6). The net solid structure with lowest energy and minimum atomic redistribution is platelike, with alternating layers of the two phases. Such eutectic microstructure, which appears in Figure 5.14, is quite similar to pearlite formed by the eutectoid transformation we discussed in Section 4.5.

What happens to the microstructure when we have compositions richer in either element? The inverse lever rule is very helpful in figuring out the resulting

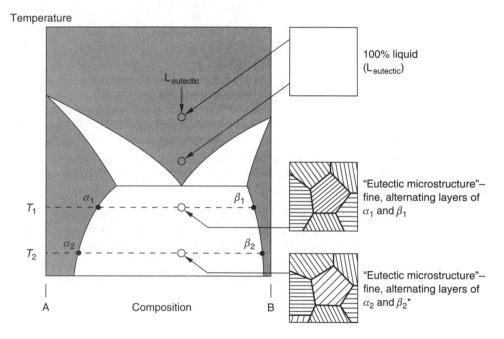

*The only differences from the T_1 microstructure are the phase compositions and the relative amounts of each phase. For example, the amount of β will be proportional to:

$$\frac{\chi_{\text{eutectic}} - \chi_\alpha}{\chi_\beta - \chi_\alpha}$$

Figure 5.14
Microstructure developed during solidification of a eutectic alloy
(From J. F. Shackleford, *Introduction to Materials Science for Engineers*, 3rd ed., Macmillan, 1992.)

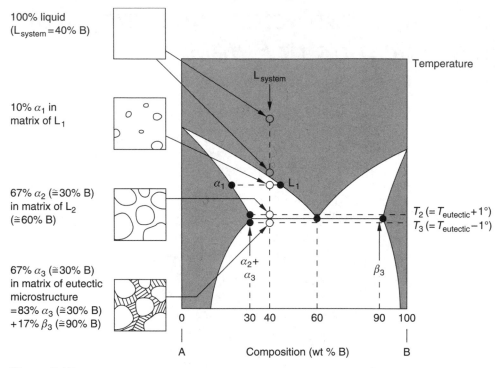

Figure 5.15
Microstructure developed during solidification of a hypoeutectic alloy
(From J. F. Shackleford, *Introduction to Materials Science for Engineers*, 3rd ed.,
Macmillan, 1992.)

microstructure. If we have a hypoeutectic composition (richer in the primary component) such as depicted in Figure 5.15, we use the lever defined by α and the eutectic composition. The microstructure is in a matrix of eutectic structure α. For a hypereutectoid alloy, we use the lever defined by the eutectic composition and the ß composition. The resulting microstructure is in a matrix of eutectic microstructure ß, as depicted in Figure 5.16.

Summary

Metallic alloys are prepared in the molten state, either during refining or by melting pure metal mixtures. Structural steel is made in a blast furnace and refined further in a basic oxygen furnace. Ferrous metals are also produced by electric arc furnaces and induction furnaces. Nonferrous alloys can be melted in induction furnaces or in fuel-fired crucible furnaces. Vacuum refining is accomplished by adding vacuum capabilities to induction furnaces and by consumable electrode remelting processes. Metal alloys can be shaped by pouring molten metal into molds and allowing the liq-

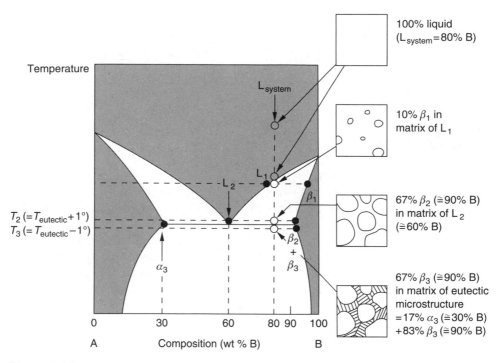

Figure 5.16
Microstructure developed during solidification of a hypereutectic alloy
(From J. F. Shackleford, *Introduction to Materials Science for Engineers*, 3rd ed.,
Macmillan, 1992.)

uid to solidify, which forms the basis of the foundry industry. Solidification produces both macroporosity and microporosity in addition to segregation and cast ingot microstructure that affect the properties of the cast metal alloys — usually adversely! Solidification of eutectic alloys results in microstructures peculiar to the eutectic composition and the eutectic microstructure combined with the separate phases, with amounts determined by the lever rule.

Terms to Remember

basic oxygen furnace	electric arc furnace
blast furnace	equiaxed grains
chill zone	fuel-fired furnace
columnar grains	grain growth
coring	induction furnace
crucible	ingot
distribution coefficient	macroporosity

microporosity   shrinkage

nucleation   solidification

nonmetallic inclusion   steelmaking

riser   vacuum melting

segregation

Problems

1. Describe two ways of making steel.
2. Why are Ni-Fe magnetic alloys made by induction melting pure Ni and pure Fe, then casting the molten metal into an ingot, and remelting by vacuum consumable arc?
3. Aluminum alloys are frequently melted in fuel-fired crucibles. Before casting, the molten metal is degassed. Give an explanation for this procedure.
4. Explain the shape of the cast microstructures in Figure 5.7.
5. How and why does microporosity affect mechanical properties?
6. Explain how segregation occurs.

6

Heat Treatment of Metals and Alloys

In Chapter 4, we were introduced to equilibrium phase diagrams that describe the phase relationships after infinite time has been allowed to reach the lowest system energy. In Chapter 5, we learned that solidification produces nonuniform chemistry and grain structures that are not characteristic of equilibrium. We have been welcomed to the real world and now must learn how to make useful metals and alloys. In this chapter, we are going to learn about **diffusion**, a chemical redistribution of atoms in a solid, and then examine heat treating, which takes advantage of diffusion.

6.1 Atomic Diffusion

In order to change the properties of metals and alloys, we have to move atoms around within the crystalline solid. We can do this either by heat treatment at elevated temperatures or by mechanical deformation. Mechanical deformation will be covered in Chapter 9, so discussion here will be confined to heat treating. Atom movement in heat treating would not be possible in perfect crystals, but we have learned that there are defects, particularly vacancies, in the crystal lattice (Chapter 3). These vacancies play an important role in the movement of atoms.

Consider a plane of atoms, such as that depicted in Figure 6.1a. We must provide energy to move atom A from its current position into the vacancy. The energy to move it, shown in Figure 6.1b, goes through a maximum that we refer to as the **activation energy** for diffusion. Remember that we want to establish the lowest energy state, or equilibrium, so there is always going to be a driving force for atom movement that will tend to eliminate segregation. The activation energy must be overcome in order to reach equilibrium, which really means that we have to raise the temperature. Therefore, **heat treating** is simply moving atoms by diffusion in the solid state by heating to high temperatures.

How fast the atoms can travel through a lattice depends on the diffusing atom size, the host crystal structure, grain size, and other factors, including activation energy. We use the term *diffusivity* to describe how fast atoms move through the lattice. Figure 6.2 shows that we must heat-treat at higher temperatures to have extensive diffusion, so it is important to remember that the two most important factors in heat treatment are time and temperature.

6.2   *Heat Treatment of Cast Metals*

Properties of cast metals are adversely affected by microporosity, segregation, and microstructural differences. We cannot change microporosity by heat treatment, but

Figure 6.1
Activation energy to move an atom into a vacancy: (a) vacancy within an atomic plane, (b) potential energy of atom A moving into a vacancy position. The activation energy is the difference between the maximum and minimum energy.

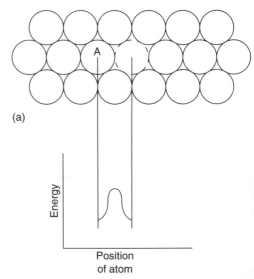

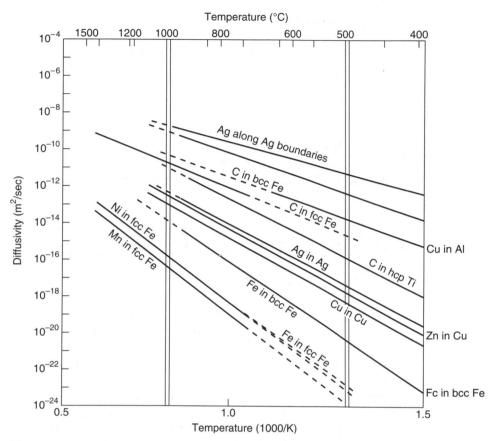

Figure 6.2
Diffusivity for selected metal alloys
(After L. H. Van Vlack, *Elements of Materials Science and Engineering*, 6th ed.,
Addison-Wesley, 1989.)

we can reduce segregation and, in some cases, affect the microstructure by heat treatment.

The most basic heat treatment is called **homogenization** because it is intended to reduce or eliminate the segregation. In this procedure, cast metals are heated to temperatures just below the solidus, then soaked for long periods of time in order to approach equilibrium conditions. The length of time required depends on the temperature, ingot size, and many other factors, but can be anywhere from 8 to 18 hours for mild steel in a soaking pit.

We can also affect microstructure by heat treatment. Cast metals have grain boundaries that form during solidification. These grain boundary surfaces have energy associated with them that is inversely dependent on the grain diameter — the smaller the diameter, the higher the surface energy! Therefore, energy will be lowered if the grains grow in size. After a soaking homogenization, grain size is large.

Other than affecting grain growth, however, we do not change the cast microstructure unless there is a solid-state phase transformation, such as in the Fe-Fe$_3$C eutectoid reaction. We will examine steel heat treatment in much more detail later in this chapter, but use the eutectoid reaction here to demonstrate that the cast microstructure can be altered by heat treating in the fcc austenite region, then cooling through the eutectoid temperature. Just as in solidification, grains of the bcc ferrite and cementite are nucleated from the decomposition of the fcc austenite. Of course, the opposite is also true upon heating through the eutectoid temperature, with ferrite and carbide dissolving and nucleating austenite. Thus we can change and control the grain structure of castings where solid-state transformations occur. (Note that homogenization also takes place by heating into the austenite.)

6.3   Furnaces and Atmospheres for Heat Treatment

Heat-treating furnaces come in many sizes, from small laboratory models to very large furnaces capable of heating many tons of metal alloys. These furnaces can be heated electrically or by fuel and can be designed for batch heating or designed for conveyors passing the metal alloys through the hot zone. In some cases, heat transfer to the workpieces is improved by immersing them into molten salts. Cost of the furnaces of course depends on the size and type but also is affected by the maximum temperature it is designed for. Higher temperatures require better insulation and different support materials for the workpieces. Controlled environments add significant cost, but frequently we can introduce *retorts*, which are sealed containers that fit into the furnace chamber.

Most metals oxidize when heated. In many metals, such as steel, this is no problem because the oxides are easily removed after heat treatment with no adverse effects. Other metals, however, require protection during heat treatment. For example, removing oxide from copper wire would be difficult because of its large surface area. If it were not removed, then further drawing would result in poor surface finish because of the abrasive oxide. The only answer, then, is to protect copper wire by heating it in an inert gas such as nitrogen, argon, or helium. In some instances, a vacuum is required to prevent oxidation.

We can also employ **furnace atmospheres** that react chemically to purify our materials. For example, the Fe-Ni alloys of Figure 3.3 were heat-treated in very dry hydrogen at 1200°C to remove the sulfur. Hydrogen is also used in the heat treatment of electrical steels (which are Fe-Si alloys, not Fe-C alloys), and there is moisture present as well to promote decarburization during the annealing process. There are also other chemical reactions that occur in furnaces, which we will discuss later in the chapter.

Case Study 6.1

Poor Selection of Heat-Treating Atmosphere

Superconductivity, the absence of electrical resistance, has been both a laboratory curiosity and a commercial interest since its discovery in 1916. There is a superconducting magnet that operates in liquid helium in every magnetic resonance imaging (MRI) machine used in the field of medicine. A laboratory goal has been to produce superconductors that have higher transition temperatures, so that costly liquid helium can be eliminated. One attempt was to produce wire containing niobium filaments in a bronze matrix, then wind the wire into a solenoid. The solenoid was to be heated at high temperatures to diffuse tin from the bronze in order to react with the niobium, thus converting the niobium and tin into Nb_3Sn, a very brittle superconductor. In order to protect the tin from diffusing to the surface, a tantalum diffusion barrier separated the Nb/bronze from the outer copper surface. The bronze work-hardens by the deformation in wire drawing, so heat treatment was frequently necessary to soften the wire (work hardening and process annealing are covered in Chapter 9).

MC Superconductors did not have heat-treating facilities for this project and the rod size for the first process anneal was too large for their normal heat-treating vendor, Vac-Treat, Inc., so a substitute vendor, HT Corporation, was approached. Because their vacuum furnace system was contaminated, the anneal was performed in a hydrogen furnace. This procedure was repeated successfully until the wire had been reduced to 0.040 inch diameter. When the wire was returned after annealing, however, it kept breaking in the first die.

After many discussions and futile wire-drawing attempts, the fractured ends were sent to a laboratory for examination in a scanning electron microscope. The answer to the problem was immediately apparent. As Figure 6.3 illustrates, the hydrogen had diffused through the copper sheath and embrittled the bcc tantalum. The solution was equally apparent. Now that the wire size was small enough to spool, it could fit into the VacTreat furnace. Hydrogen was removed by the vacuum and the wire was successfully drawn to its final size upon return to MC Superconductors.

6.4 The Decomposition of Austenite ___

In Chapter 4, we looked at equilibrium cooling curves that required thermal arrests whenever there was an invariant reaction. We did not, however, examine the length of time necessary for a phase to disappear and permit cooling to continue. Now that we are in the real world, though, we must determine how long transformations take.

Figure 6.3
Fracture surface of (*left*)
Nb/bronze, (*center*) tantalum,
and (*right*) copper composite.
The tantalum has been
embrittled by hydrogen.

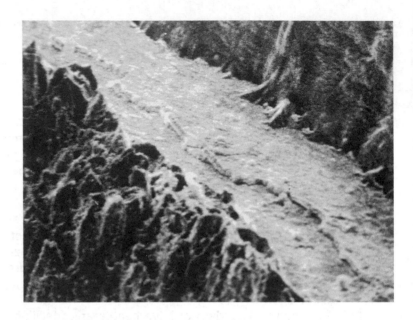

The time is necessary for diffusion to occur, whether on heating or on cooling. Our starting material should be uniform — that means we want to have a single-phase solid solution. In the Fe-Fe₃C phase diagram, it also means we must heat into the austenite region, a process called **austenitization**. The eutectoid portion of the Fe-Fe₃C phase diagram, reproduced in Figure 6.4, shows that the actual temperature decreases with carbon content up to the eutectoid composition, then increases for higher carbon content up to the eutectic temperature. Typically, we heat to about 100°F above these temperatures and allow about one hour per inch of thickness.

Figure 6.4
The eutectoid reaction in the
Fe-Fe₃C phase diagram

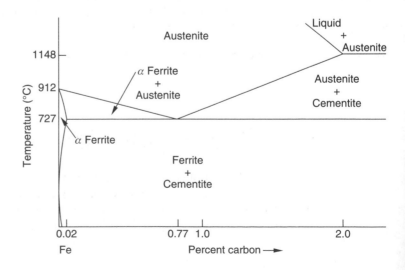

Decomposition upon cooling from the austenite temperatures will form the same type of microstructure as that for solidification of a eutectic alloy, discussed in Chapter 5. For compositions less than 0.77% C, the eutectoid composition, the microstructure will consist of ferrite plus the eutectoid microstructure, alternating platelets of Fe_3C and ferrite, better known as pearlite. Figure 6.5 illustrates the microstructure of fine pearlite. For compositions greater than 0.77% C, the microstructure will be pearlite and cementite.

The time element, however, is our prime concern. Most of our understanding has come from studies of **isothermal** transformation of samples **quenched**, that is, cooled rapidly, from the austenite phase to specific constant temperatures, then held until decomposition takes place. Figure 6.6 illustrates how the data can be summarized on an isothermal transformation diagram, or **TTT (time-temperature-transformation) diagram**. The shape of this diagram results from a diffusion-limited transformation at low temperatures and an energy-limited transformation at high temperatures. At high temperatures where diffusion occurs fastest, the pearlite that is formed is coarse. Decomposition occurs and the cementite platelets as well as the ferritic spacing between platelets become large, producing coarse pearlite. Diffusion becomes slower at lower temperatures and the fine pearlite of Figure 6.5 can be

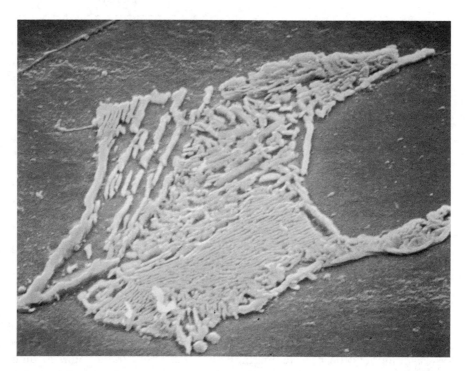

Figure 6.5
Microstructure of fine pearlite (15,000×)

Figure 6.6
Isothermal transformation (TTT) diagram for eutectoid steel

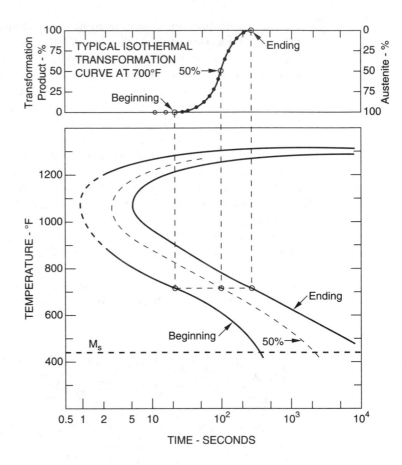

obtained only for transformation near the "knee" of the TTT diagram. When transformation occurs below the knee, a feathery microstructure called **bainite** is formed (see Figure 6.7).

When we cool the steel rapidly enough to miss the knee of the TTT diagram, the driving force for transformation is so strong that the fcc austenite transforms by a shearing of the crystal lattice, forming a bcc tetragonal structure that we call **martensite**. Because the transformation is by shear, there is no diffusion and the carbon atoms that were dissolved in the austenite are now in a supersaturated, distorted (noncubic) solid solution. The net effect is martensite, a very hard, strong, and brittle material. Hardness of the martensite depends on the carbon content, as we might anticipate.

Martensite is practically useless as formed, but it can be softened by reheating, or **tempering**, at temperatures below the eutectoid temperature. This treatment can produce a very strong, tough material that we call *tempered martensite*. Properties depend on both tempering temperature and the amount of time at that temperature. Because of the control of properties and the strengths that can be achieved at little cost, quench-and-temper heat treatments are the most common steel heat treatments for optimum combinations of strength and ductility.

Figure 6.7
(a) Microstructure of upper bainite formed by a complete transformation of a eutectoid steel at 450°C (850°F).
(b) Microstructure of lower bainite formed by a complete transformation of a eutectoid steel at 260°C (500°F). The white particles are Fe_3C, and the dark matrix is ferrite. (Electron micrographs, replica-type; magnification 5000×) (From H. E. McGannon (ed.), *The Making, Shaping and Treating of Steel,* 10th ed., Association of Iron and Steel Engineers, 1986.)

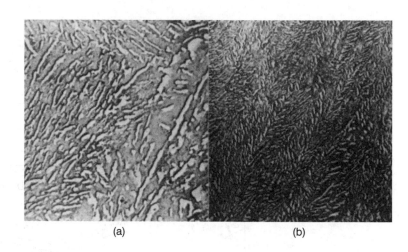

(a)　　　　　　　　(b)

6.4.1　Hardenability

A subject closely related to the TTT diagrams is the hardenability of steel. This term differs from hardness of steel because **hardenability** is a measure of the *uniformity of hardness* through thick sections of steel. We can understand this quite easily by considering the hardness at the surface and center of a thick steel bar. Martensite can be formed at the surface, but at the interior, slower cooling (limited by heat transfer through the steel thickness) produces transformation to the softer ferrite and pearlite structure. Hardness is therefore not uniform through the thickness of the bar. Hardenability is measured by end-quenching a sample (removing heat only through the end), then measuring the hardness along the length of the sample.

In order to understand hardenability, however, we cannot look only at the TTT diagrams because the transformation occurs on continuous cooling, not isothermally. Therefore, we construct a continuous cooling curve by superimposing cooling curves for a particular location onto the TTT diagram to determine what microstructure will be formed by the transformation at that specific location. Such simple correlation between the continuous cooling curve and the TTT diagram is shown in Figure 6.8.

6.4.2　Alloying Effects on Transformation

A major concern in heat treatment is that the cooling rate cannot avoid the knee of the TTT diagram, thus prohibiting the formation of martensite. The knee of the TTT diagram, however, can be shifted to longer times by increasing the carbon content, by adding small amounts of alloying elements such as Mn, Cr, Ni, Si, Mo, and others, and by starting with larger austenite grain size. Small alloy additions are pre-

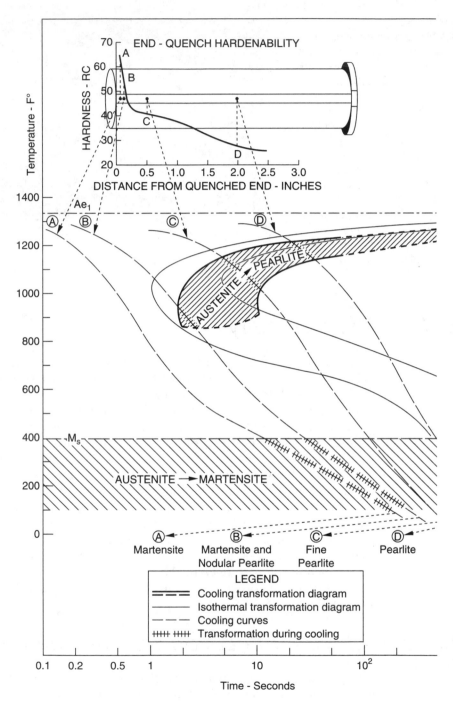

Figure 6.8
Superposition of continuous cooling on TTT diagram for eutectoid carbon steel

ferred, forming a group of alloys known as high-strength low-alloy, **HSLA, steels**. By adding alloy elements and shifting the knee of the TTT diagram to longer times, the hardness of the interior is raised. The overall hardness of the bar is not only higher, but more uniform as well. Martensite that is formed through much or all of the section can then be tempered to uniform desired hardness.

Figure 6.9 shows the TTT diagram for one common HSLA steel, AISI/SAE 4340, which contains Ni, Cr, and Mo additions combined with about 0.40% C. In this alloy, the alloying effect on diffusion limits the transformation at higher temperatures so much that a double knee is formed. As the figure shows, though, quenching can produce martensite at very large depths, assuring optimum hardenability by quench-and-temper heat treatment.

6.4.3 Heat Treatment of Steel

Heat treatment that depends on the decomposition of austenite is used for controlled mechanical strength and ductility. There are specific names that describe the actual heat-treating practice. For example, holding for very long times just below the A_1 temperature tends to break up the coarse pearlite into cementite spheres in a ferrite matrix. We call this type of treatment, rare as it is, **spheroidization**. When we austenitize, then cool slowly to form soft, coarse pearlite for easier machining, we call it a **full anneal**. This is in contrast to process anneal, which we cover in Chapter 9, and stress relief anneal, discussed in the next section. When we austenitize, then air cool, we call the treatment **normalizing**. When we cool from the austenite and miss the knee of the curve but stop short of martensitic transformation, we form bainite isothermally; this we call **austempering**.

The most common heat treatment, however, is the quench and temper. Quenching produces martensite, which of course is hard and brittle. Tempering, however, is heating at temperatures in the range of 400°F to 1200°F. The martensite slowly transforms to ferrite and cementite, but the microstructure characteristic of tempered martensite is very different from annealed ferrite and cementite. Such heat-treating practices allow us to control the strength and ductility of the steels.

Direct quenching to martensite can be harsh, even causing cracking because of the stresses induced by the sudden expansion and brittle characteristics of the martensite. We can reduce the severity by quenching below the knee, slow cooling further to form martensite, and then tempering. We call this process **martempering**. Figure 6.10 summarizes the major heat-treating procedures that involve eutectoid decomposition.

6.4.4 Quenching Media

During any quenching process, we must be concerned with the heat removal rate in order to avoid transformation before cooling below the knee of the TTT diagram. For steels containing substantial alloys, such as the HSLA steels, this is relatively

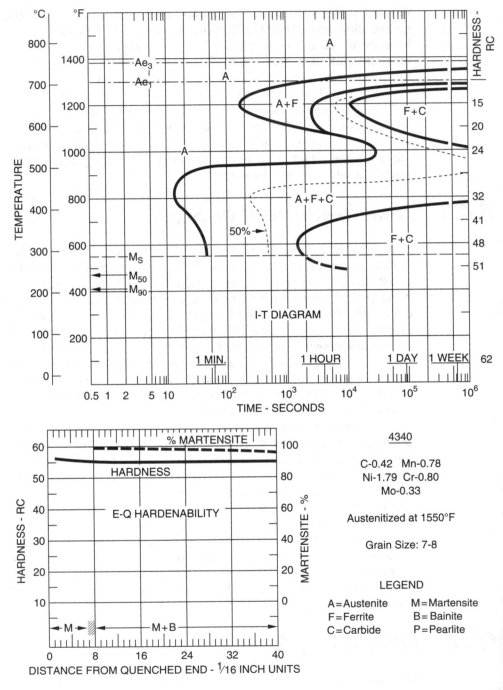

Figure 6.9
TTT diagram for AISI/SAE 4340 high-strength low-alloy steel

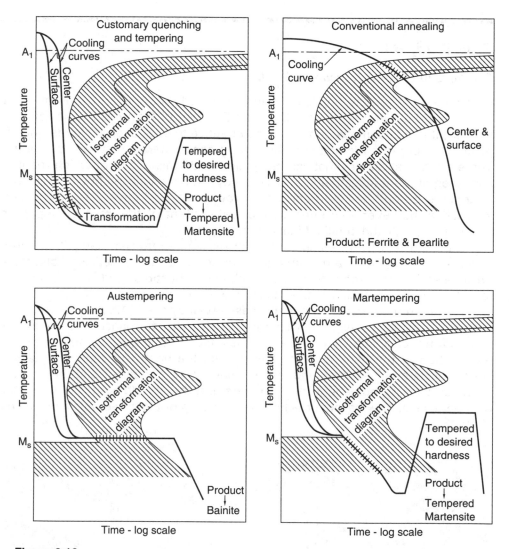

Figure 6.10
Summary of heat-treating procedures and products produced by decomposition of austenite

easy. Such alloys are usually quenched in oil. For alloys that must be quenched much faster, however, we must pay close attention to the heat removal rate. The cooling rate can be increased by lowering the temperature of the quenching medium, increasing the conductivity of the medium, and by agitating the medium. Brine, for example, cools faster than plain water, and iced brine even faster. If the medium is not agitated, it boils at the metal surface and heat transfer is slowed considerably.

Case Study 6.2

Problems in Finding Treasure in Trash

Public attention to recycling is justifiably deserved, but only a fraction of our trash can be recycled. What can we do with the rest? Because of high costs associated with sanitary landfills, it has become economically competitive to seek alternative methods for trash disposal. One popular method is to burn the trash, generating steam to produce electricity. There are two strategies that can be applied to do this, either of which requires expensive capital equipment investment. In Case Study 11.4, we will see how one alternative, where all trash is burned, leads to corrosion problems. The I Love Trash Corp. uses a second alternative, however, that enriches the fuel content of the trash by separating it before incineration; only the enriched fuel is burned.

I Love Trash Corp. places the trash on conveyors that carry it to shredding mills, which are simply very large hammers that compact and shred the solid trash. Noncombustible solids are removed and the enriched combustible solids are compacted and transferred to a separate facility where steam and electricity are generated.

The hammer parts are made of a cast low-alloy steel (AISI/SAE 8630), purchase specifications for which required only chemical analysis conformance on the part of the foundry. The hammer arm, shown in Figure 6.11, is attached to a hollow-ended hammer and transfers motion to the hammer. With this design, the hammers can be replaced when they wear, but hammer arms should last indefinitely. A number of hammer arms have failed, however, causing damage to the hammers and the shredding mill and creating long production delays. The reason for failure was cracking of the cored hole, such as that shown in Figure 6.12.

Figure 6.11
Cast AISI/SAE 8630 steel hammer arm for a shredder

Figure 6.12
Cracking in the cored hole of the hammer arm (left hole in Figure 6.11)

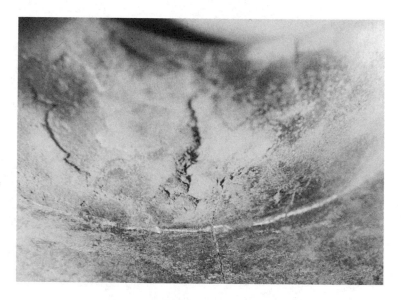

Because of the nature of the work stresses, cracking was thought to be caused by fatigue, but the fracture surfaces were too corroded to examine. The cracking patterns, however, were consistent with tensile fatigue failure in a pin connector. Chemical analysis showed conformance to that specified for the low-alloy steel. Mechanical properties were measured and appear as data points on Figure 6.13, which shows the published information on tempering effects on properties of AISI/SAE 8630 steel. Small samples were again heat-treated and hardness values were measured, yielding results consistent with this figure.

It was concluded that the failure could have been avoided if the hammer arms were tempered at 800°F. Purchase specifications were changed to include the tempering temperature and mechanical properties as well as the chemistry.

6.4.5 Surface Hardening

Most of the time, we attempt to prevent carbon loss by surface oxidation during heat treatment. There are times, however, when we want to have the opposite effect, a hard surface to provide wear resistance and a ductile inner core to provide toughness. Examples of applications for surface hardened steels are gears, sheaves, bushings, and rolls. We have learned how such selective **surface hardening** might be produced. One method is to austenitize only the surface, then quench to form martensite. This is accomplished by induction heating or by flame or laser hardening and is usually confined to steels containing sufficient carbon, 0.40% and higher, to form hard martensite.

Figure 6.13
Comparison of properties of the failed hammer arm with typical tempered properties of AISI/SAE 8630 steel

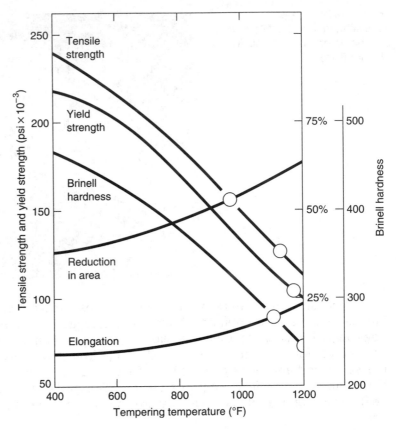

Carbon can also be diffused into the surface by packing the workpiece in charcoal or graphite, then heating for very long time periods, a process known as **carburizing**. Or it can be formed by reaction with gases such as methane (natural gas) or by immersion in liquid carburizing salts. In all cases, quenching from the carburizing temperatures is necessary.

Nitrogen can also provide hardened surfaces, but only for special steel alloys (nitralloys) containing small amounts of aluminum, chromium, molybdenum, vanadium, or tungsten. The advantage of **nitriding**, producing a hard surface on a metal by introducing nitrogen, is lower processing temperatures and elimination of quenching.

Case Study 6.3

A Quenching Problem

Some railroad freight car suspension systems utilize a snubber (shock absorber) assembly that includes inner coil springs and snubbers that are inserted into a snubber sleeve and outer coil spring. Figure 6.14 illustrates the snubber sleeve and outer coil.

The coil spring is hot-coiled from AISI/SAE 8655 steel that is austenitized at 1550°F, oil quenched, and tempered. The snubber sleeve is welded using forged caps and seamless tubing made of AISI/SAE 1020 steel. Specifications call for the weld and cap to be coated with "Non-Case" to prevent hardening and the sleeve to be pack-carburized for 72 hours at 1650°F, then water quenched to a hardness of Rockwell C 45–50. In order to accomplish this for the inside as well as the outside of the sleeve, a spray quench that utilizes pond water has always been employed.

Quality control procedures revealed that the hardness requirement was not met during the first week of July, causing a backup in assembly and unmet shipping plans. Fast response was needed, but the long carburizing time limited quick resolution. Metallography showed case depth to be adequate, so the quenching practice was altered. Using cut sections, specified hardness was achieved for water, brine, and iced brine quenching, but the spray-quenched sleeves still failed. The temperature of the pond water was checked and found to be 80°F. Although no records of pond temperature had been maintained, block ice was added to reduce the pond temperature to 60°F before the next batch was quenched. This strategy proved adequate to meet specifications.

6.4.6   Stress Relief Annealing

There is another effect of heat treatment that cannot be neglected because of the speed of heating or cooling. Limited heat transfer from the interior to the surface

Figure 6.14
Snubber assembly for a railroad freight car suspension system: snubber sleeve (left) and coil spring (right)

can create thermal stresses during rapid heating or cooling by quenching that can lead to distortions and even fracture. Thermal stresses can be eliminated by **stress relief anneals**, heating for short times at temperatures below the A₁ temperature.

6.5   Precipitation Hardening, or Age Hardening

Practical heat treatment involving decomposition of austenite is limited to ferrous metals, but strengthening by precipitation of a second phase occurs in a number of alloy systems, including both ferrous and nonferrous alloys. Strengthening actually results by preventing dislocations from moving when stress is applied, but we will learn more about this important subject in Chapter 9. Here, we are only interested in the process of heat treating to strengthen.

Our understanding of **precipitation hardening**, or **age hardening**, has primarily used the model system of the aluminum-rich end of the Al-Cu phase diagram, shown in Figure 6.15. If we select an alloy containing 5% Cu or less, we can heat it into the single-phase κ region, a process called **solutionizing**. If we were to cool this, the κ would decompose slowly to the two-phase $\kappa + \theta$, much the same way that austenite decomposes. However, if we quench the single solid solution, there is no time for diffusion to occur and there is no diffusionless transformation. The single phase is supersaturated with copper. If we reheat the supersaturated solid solution but stay below the solvus, the equilibrium θ-phase gradually forms. However, the hardness increases to a maximum, then decreases when the equilibrium conditions are met!

We now know that the maximum hardness is associated with the beginning of the nucleation and growth of the second phase. The nuclei that form are clusters of copper atoms only a few atoms thick, but even these increase the strength. As the nuclei develop (see Figure 6.16), they become disc shaped with coherent features; that is, the atoms of the adjacent phases are aligned one-to-one. This is true for the face of the disc, but through the thickness of the disc the atoms are incoherent, that is, not aligned. As development continues, the coherency is partially lost at first, corresponding to what we call θ', then the final equilibrium θ incoherent phase is formed. We call the coherent phases **GP zones** (GP-1 and GP-2) after Guinier and Preston, the researchers who first discovered the phenomenon. Maximum hardness is associated with the GP-2 and θ' precipitates, and once the maximum is surpassed, we refer to the reduced hardness as overaging.

The age hardening is usually done commercially by solutionizing in the single-phase solid solution, quenching to form the supersaturated solid solution, then aging at low temperatures for long time periods to ensure uniform hardness throughout all sections. The aging curves for 6061 aluminum alloy are shown in Figure 6.17. This alloy contains small amounts of silicon, magnesium, and copper, all of which form eutectic alloys with aluminum.

Figure 6.15

The aluminum-rich end of the Al-Cu phase diagram

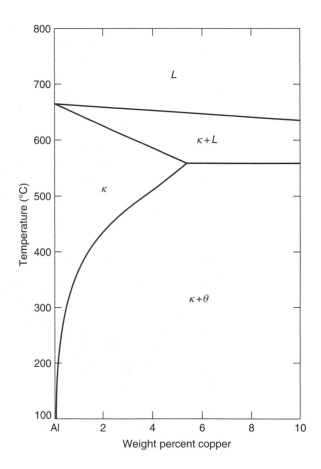

Age hardened alloys have standard temper designations to denote the aging heat treatment. The temper designation for as-fabricated alloys is F, that for annealed alloys is O, and that for work hardened metals is H, but for age hardening the designations are

T1 Naturally aged to a stable hardness after shaping at an elevated temperature

T3 Solution treated and cold worked, then naturally aged to a stable hardness

T4 Solution treated and naturally aged to a stable hardness

T5 Shaped at an elevated temperature, cooled, and naturally aged to a stable hardness

T6 Solution treated, quenched, and artificially aged

T8 Solution treated, cold worked, and artificially aged

Note that cold work that strengthens metals (see Chapter 9) can be combined with the precipitation hardening heat treatments.

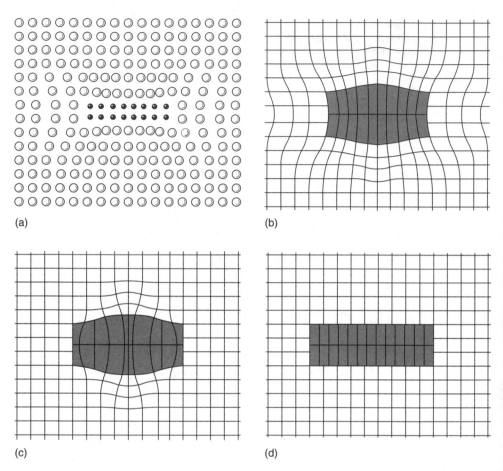

Figure 6.16
Stages in the development of the θ-phase in aging of Al-Cu supersaturated solid solutions: (a) GP-1 zone, (b) GP-2 zone, (c) θ' structure, (d) equilibrium θ structure (From W. Hayden, W. G. Moffatt, and J. Wulff, *The Structure and Properties of Materials,* Vol. 3: *Mechanical Behavior,* John Wiley & Sons, 1965.)

Summary

Properties of metals and alloys can be changed significantly by heating to high temperatures and controlling decomposition during cooling. At high temperatures, composition is altered by atomic movement (diffusion), making time and temperature the two most important factors in heat treatment. Cast metals are homogenized at the highest heat-treating temperatures to reduce segregation. Properties can be controlled in steels by heating into the single-phase austenite region, then controlling the decomposition during cooling. Time-temperature-transformation studies are useful in selecting the right heat treatment. Full annealing is slow cooling from austenite to soften for machining. Normalizing is an air cool and forms a somewhat harder

Figure 6.17
Aging of 6061 aluminum alloy

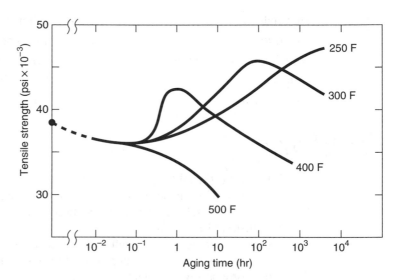

microstructure. Slower cooling promotes coarse pearlite, whereas decomposition to pearlite at the knee of TTT diagrams produces fine pearlite. Transformation below the knee of TTT diagrams produces bainite. Rapid cooling or quenching leaves no time for diffusion to occur, and the hard microstructure formed is martensite. Martensite can be reheated (tempered) to produce a strong, tough steel with reasonable ductility. Hardenability is the capability of developing hardness through larger sections, not only at the surface; it is benefited by alloying to form HSLA steels. Surface hardening is also useful in many instances for wear resistance. We can harden the surface mainly by adding carbon at the surface or by induction heating the surface only and quenching to form martensite. Precipitation hardening, or age hardening, is practiced in nonferrous alloys for strengthening. Systems that can age harden have increased solubility as temperature is increased and alloys must be heated into a single-phase region. Quenching produces a supersaturated solid solution that can be age hardened by diffusion and decomposition at intermediate temperatures.

Terms to Remember

activation energy	GP-zones
age hardening	hardenability
austempering	heat treating
austenitization	homogenization
bainite	HSLA steels
carburizing	isothermal
diffusion	martempering
full anneal	martensite
furnace atmospheres	nitriding

normalizing   spheroidization

pearlite   stress relief anneal

precipitation hardening   surface hardening

quench   tempering

solutionizing   TTT diagram

Problems

1. What is diffusion?
2. The role of vacancies in atomic diffusion was explained in this chapter. Do you think other defects such as dislocations and grain boundaries might also affect diffusion? Explain.
3. Ingots are deformed at high temperatures into billets for plate or sheet product. Do you think it is as important to homogenize an ingot for this purpose as it is for a casting? Explain.
4. Explain the importance of steel heat-treating atmospheres when we want to remove carbon or increase carbon in the steel surface.
5. Explain the microstructures of pearlite and bainite and compare them to that of tempered martensite.
6. Thin strips of eutectoid steel are heat-treated as follows:
   a. heat 1 hr at 1500°F; quench in water
   b. heat 1 hr at 1500°F; water quench; reheat 1 hr at 600°F
   c. heat 1 hr at 1500°F; quench in molten salt at 500°F and hold 1 min; quench in water
   d. heat 1 hr at 1500°F; quench in molten salt at 600°F and hold 20 min; quench in water
   e. heat 1 hr at 1500°F; water quench; reheat 1 hour at 1200°F

   In each case, describe the microstructure following heat treatment (use Figure 6.6).
7. List the effective heat removal rates for the following quenching media:
   a. water
   b. iced water
   c. brine
   d. iced brine
   e. oil

   What is the effect of agitation in each medium listed?
8. Explain what surface hardening is, how it can be accomplished, and an application it might be used for.
9. What features of a phase diagram are necessary in order for precipitation hardening to apply?
10. In precipitation hardening alloys, why does maximum hardness occur before the equilibrium second phase is completely formed?

7

Ferrous Alloys

Metals and alloys can be described by their major constituent, such as iron-base alloys and copper-base alloys, or they can be referred to more generally as ferrous or nonferrous alloys. The latter classification is commonly used because of the broad acceptance of an iron and steel industry, separate from all other alloys. Ferrous alloys can be defined as alloys that contain iron as the main element; nonferrous alloys can be defined as those alloys whose main element is any metal other than iron. In Chapters 7 and 8, keeping this definition in mind, we will look at both ferrous and nonferrous alloys with only a few exceptions. For example, we will consider Invar (Fe-36% Ni) and Fe-50% Ni magnetic alloys in Chapter 8, because their behavior is defined more by the influence of the nickel. On the other hand, here we include some alnico alloys that have less than 50% Fe because the behavior of this series of permanent magnetic alloys is predominantly ferrous.

7.1   Steel

We normally think of steel as an alloy of iron and carbon, but this interpretation is incorrect. Stainless steel and electrical steel contain little or no carbon and we will see that cast iron can be considered as a steel matrix containing graphite. In this section, we will only consider steels that derive their properties in large part from the iron base and the carbon content. Descriptions of standard alloys begin with standards based upon composition developed by cooperation of the American Iron and Steel Institute (AISI) and Society of Automotive Engineers (SAE), which are reproduced in Table 7.1. There are other standards, such as those of the American Society for Testing and Materials (**ASTM**) and the American Petroleum Institute (**API**), but the **AISI/SAE** standards are the most frequently used.

The most common steels are the **plain carbon steels**, or **mild steels**, that are used for many structural, sheet metal, and tensile applications. For example, sheet metal for ducts and metal cabinets, nails and screws, and concrete-reinforcing rods are normally made of low-carbon steel. Structural beams and steel plates used in construction are mostly made from AISI/SAE 1020 steel. There are few applications for medium-carbon (0.25% C–0.50% C) because these alloys lack the necessary hardenability and therefore can only be used for relatively thin sections. Plain carbon steels with high carbon content find compressive applications as chisels and hammers and tensile applications as wires in cables. For example, elevators are suspended by cables containing high-carbon steel wires. A novel but very important use of 1080 steel wires is for pianos and violins, which gives rise to the term **music wire** for wires of this composition.

Table 7.1
AISI/SAE classification of standard alloy steels

Classification	AISI/SAE number*	Alloy additions
Carbon steels	10xx	0.5% Mn**
	11xx	1% S (free machining)
Mn steels	13xx	1.75% Mn
Ni steels	2xxx	0.5–5.0% Ni
Ni-Cr steels	3xxx	1.2–3% Ni, 0.6–1.6% Cr
Mo steels	41xx	Cr, Mo
	43xx	1.8% Ni, Cr, Mo
	46xx	Ni, Mo
Cr steels	5xxx	Cr
Cr-V steels	6xxx	Cr, V
Ni-Cr-Mo steels	8xxx	0.5% Ni, Cr, Mo

* The last two digits in the classification denote the carbon content, for example, 1020 and 8620 steels contain 0.20% C.

** One-half percent Mn is added to all steels to eliminate **hot shortness**, a brittleness caused by segregation of sulfur to form low-melting-point grain boundary films.

Case Study 7.1

The Case of the Broken Pipe

We depend on plain carbon steel pipe to carry many fuels, including oil and natural gas, through buried pipelines. Some pipe compositions must conform to ASTM A53, grade B, which contains 0.30% C and 1.20% Mn, with specified ultimate tensile strength of 60,000 psi. Excavation projects frequently uncover buried pipelines in urban areas. Although notice is required to all utility companies as well as local authorities, excavation is usually witnessed by only those working directly on the project. All too frequently, backhoes or falling rocks inadvertently strike the exposed pipeline. However, such occurrences usually cause nothing more than small indentations in the pipe.

When a relatively small indentation caused the pipe to actually fracture (see Figure 7.1), a failure analysis was conducted. The hardness showed that the steel met the specified tensile strength and the microstructure was mostly ferrite with small patches of pearlite. Yet the fracture surfaces were brittle, not ductile, as shown in Figure 7.2. Interviews indicated the ambient temperature was a little above freezing at the time of the break. Sections of the pipe were cut and impacted at several different temperatures to determine if the fracture was caused by the low temperature. Although standard size specimens could not be made, data obtained were helpful. Below about 40°F the samples broke in the same brittle fashion as the pipe had, whereas those struck sufficiently hard to break them at higher temperatures actually broke in a ductile manner. Figure 7.3 illustrates the fracture surface of a ductile failure.

Failure was correctly attributed to the **ductile-to-brittle transition** that bcc metals, including plain carbon steels, undergo when temperature is lowered. It was recommended

Figure 7.1
Deformation and fracture of subject pipe

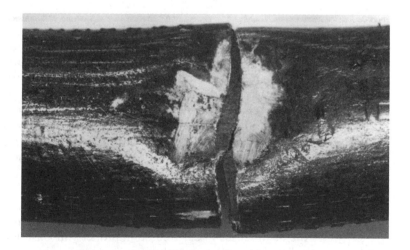

Figure 7.2
SEM micrograph of brittle fracture for the broken sample in Figure 7.1

that more training be required to alert backhoe operators to the brittle behavior of these metals at temperatures actually above freezing. The suggestion was also made that temperature be considered a major factor in flexible scheduling, except in emergency situations.

Plain carbon steels are used wherever possible for obvious economic reasons. However, we do encounter limitations that force other selections. These alloys cannot be strengthened beyond about 100,000 psi without being severely embrittled. They are not hardenable, so their thickness is very limited wherever higher strengths are necessary. Where they can be hardened, rapid quenching is necessary to miss the knee and form martensite; this leads to distortions and even cracking from induced thermal stresses. These alloys are also restricted for low-temperature applications because of the ductile-to-brittle transition behavior covered in Case Study 7.1. Last but not least, the plain carbon steels do not resist corrosion, a severe restriction (as we shall see in Chapter 11).

Low-alloy steels are heat treatable and are used for many applications where hardenability is an important characteristic. High-strength machine parts, such as the hammer arm heat-treated improperly in Case History 6.2, gears, shafts, leaf and coil springs, and many other applications for these heat treatable alloys abound.

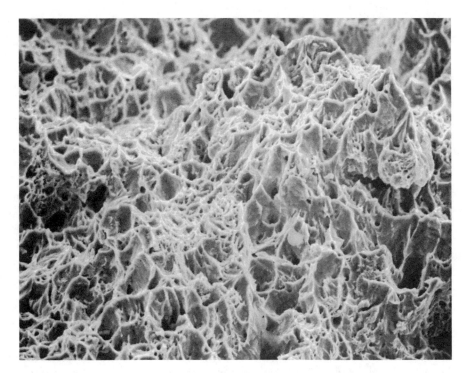

Figure 7.3
SEM micrograph of ductile fracture of steel sample with similar composition and hardness to subject pipe (1000×)

7.2 Cast Iron

Cast irons were briefly mentioned in conjunction with the Fe-Fe₃C phase diagram in Figure 4.8. These alloys are cast to shape economically because of their low melting temperatures in comparison to steel. They have a wide range of strengths and are readily machinable, but are usually brittle and have low impact strength. Most commercial cast irons contain 2–4% carbon plus silicon and manganese. They are classified, however, by the color of a fresh fracture and by the shape of the graphite present in the microstructure. The types of cast iron are

> gray
>
> white
>
> malleable
>
> ductile

Silicon, in amounts of 0.5–3%, acts as a substitute for carbon, thus altering the eutectic reaction. It also tends to promote formation of the real equilibrium phase, graphite, as opposed to cementite. Thus, upon cooling of cast irons, austenite and graphite are formed instead of austenite and cementite. Since the austenite (carbon being lowered by the formation of the graphite) decomposes to ferrite and cementite, we can think of cast irons as a steel matrix containing graphite.

The most common and least expensive cast iron is **gray cast iron**. Its microstructure, shown in Figure 7.4, consists of graphite in the form of flakes created during the eutectic solidification. These **graphite flakes** have little or no strength and contribute to brittleness because they act like stress concentration voids. Typical grades of gray cast iron have tensile strengths of 20,000–40,000 psi. Most applications require only low strength, such as pipe, or maintain the parts in compression, such as engine blocks or frames for machinery.

White cast iron has lower carbon and silicon content than gray, so carbon forms not as graphite, but as cementite. Further austenite decomposition forms a matrix of pearlite, giving an extremely hard microstructure. This alloy is used in applications such as railroad rails where wear resistance is essential. Both gray and white cast irons derive their names from the color of a fresh fracture surface.

White cast iron also serves as the first step in forming **malleable iron**, the third type of cast iron. By reheating to high temperatures, the cementite dissociates to form graphite and austenite, the graphite as an irregular-shaped nodule known as **temper carbon**. Malleable cast iron has strengths comparable to gray cast iron, but has reasonable ductility and excellent impact strength. Typical applications for this alloy are machined parts such as automotive components. The drawback to malleable iron unfortunately is a major one because long heat-treating times, up to 100 hours, are needed at high temperatures, thus adding significant cost to parts.

Figure 7.4
Microstructure of gray cast iron (200×)

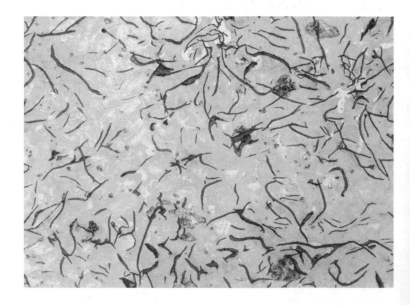

Ductile cast iron, also known as *nodular cast iron*, has a composition much like gray cast iron. Its microstructure more resembles malleable cast iron; the nodules, however, are more spherical because they are formed during solidification. This spherical shape, which is formed by adding cesium or magnesium to the liquid metal alloy just before solidification, is more stable because it has lower energy. High strength, up to 90,000 psi, is combined with good ductility, leading to typical applications of these alloys as gears, valves, and pump bodies. The cost of the additions that promote the nodule formation and strict process control makes these alloys more expensive than gray cast iron.

Table 7.2 summarizes the microstructure, properties, and composition of gray, malleable, and ductile cast irons.

7.3 Stainless Steel

As their name implies, stainless steels are used where corrosion resistance is important. We will learn more about corrosion and corrosion resistance in Chapter 11, but at this time it is only important to remember that the corrosion resistance of these alloys depends on chromium additions equal to or greater than 12%. Other characteristics, such as resistivity, magnetic behavior, and strength, are also important in the selection of specific stainless steel alloys. We generally classify stainless steels into four main categories: ferritic, austenitic, martensitic, and precipitation hardenable.

Table 7.2
Composition, microstructure, and properties of cast irons

Cast iron	Composition (%)	Microstructure	UTS* (psi)	Percent elongation
Gray	3.4 C, 2.2 Si, 0.7 Mn	Ferritic + G	26,000	—
	3.2 C, 2.0 Si, 0.7 Mn	Pearlite + G	36,000	—
	3.3 C, 2.2 Si, 0.7 Mn	Pearlite + G	42,000	—
Malleable	2.2 C, 1.2 Si	Ferrite + TC**	50,000	10
	2.4 C, 1.4 Si, 0.75 Mn	Pearlite + TC	65,000	8
	2.4 C, 1.4 Si, 0.75 Mn	Tempered martensite + TC	90,000	2
Ductile	3.5 C, 2.2 Si	Ferrite + G	60,000	18
	3.5 C, 2.2 Si	Pearlite + G	80,000	6
	3.5 C, 2.2 Si	Martensite + G	120,000	2

* UTS = ultimate tensile strength
** TC = temper carbon

Table 7.3
Composition and properties of some stainless steel alloys

Alloy no.	Type	Composition (wt %)	UTS (psi)	Percent elongation
430	Ferritic	17 Cr	75,000	25
410	Martensitic	12.5 Cr, 0.15 C	180,000	18
302	Austenitic	18 Cr, 9 Ni, 2 Mn	75,000	40
316	Austenitic	18 Cr, 12 Ni, 2 Mn	75,000	40
17-4PH	Precipitation hardenable	16 Cr, 4 Ni, 4 Cu, 0.03 Nb	190,000	14

Designations of stainless steels by the type listed by AISI are the most common. For each type, there is a general-purpose designation, with additional numbers that represent particular modifications. For example, 302 is the general-purpose austenitic alloy, 430 is the general-purpose ferritic alloy, and 410 is the general-purpose martensitic alloy. Table 7.3 lists the compositions and typical properties of these general-purpose alloys.

Ferritic stainless steel alloys are typically used for inexpensive flatware, sinks, and other general-purpose applications where hardenability is not necessary. **Martensitic** types are used where strength is important, such as for screws, nuts and bolts, machine parts, and valves. **Austenitic stainless steel** alloys, unlike the ferritic and martensitic types, are not ferromagnetic. They are more expensive because of the alloy content and because they work-harden, which increases processing costs. Their corrosion resistance, however, is superior to ferritic and martensitic types. Applications include sturdy flatware, cryogenic dewars, pressure vessels, and food- or chemical-processing equipment.

Case Study 7.2

Broken Stainless Screws

Nocorr Industries uses small socket screws to fasten components used in control equipment that must operate in a corrosive environment. These screws are made from type 410 stainless steel heat-treated to a hardness of Rockwell C (R_C) 36–40 (Brinell hardness, BHN, of 335–380); the manufacturer, however, certifies only minimum tensile strength. When screws repeatedly broke during insertion with only 16 in.-lb of torque, a metallurgical failure analysis was requested.

The simple failure analysis consisted only of examination of the fracture surfaces by scanning electron microscopy, conventional metallography, and microhardness measurements. Figure 7.5 shows the tempered martensitic microstructure. Microhardness measurements indicated that the tempering was insufficient, with equivalent hardness of R_C 43 (BHN 400). Scanning electron microscopy revealed the fracture was intergranular, occurring at prior austenite grain boundaries, with numerous microcracks, as shown in Figure 7.6.

Figure 7.5
Tempered martensitic
microstructure of screw from
lot 249 (500×)

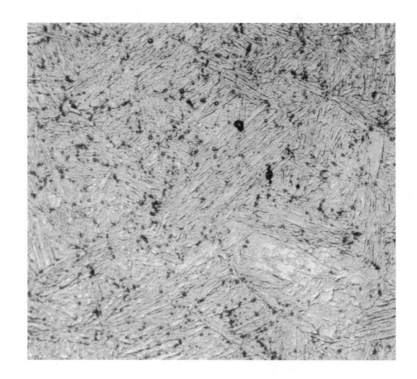

Figure 7.6
SEM micrograph of intergranular fracture at prior austenite grain boundaries (410×)

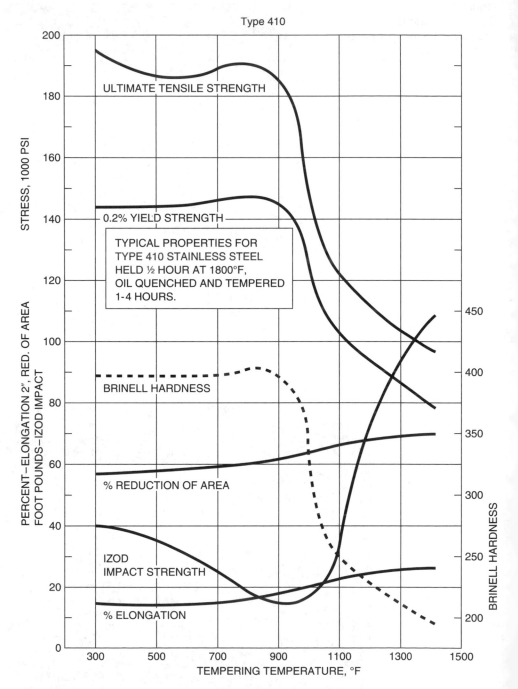

Figure 7.7
Effect of tempering temperature on tensile properties
(From *Sourcebook on Industrial Alloy and Engineering Data,* ASM International, 1978.)

By referring to manuals for heat-treating stainless steels, it was found that the martensitic alloys were probably tempered at too low a temperature in order to guarantee the minimum strength. Response to tempering of 410 stainless, reproduced in Figure 7.7, indicates that the screws were tempered at or below 900°F, whereas they should have been tempered at 950–1000°F.

The manufacturer, when confronted with the failure analysis and recommendations, was not willing to guarantee the specified hardness range, but agreed to provide the screws in the quenched condition. Nocorr developed the proper tempering treatment by using their conveyor brazing furnace, thus eliminating production assembly problems.

7.4 Tool Steel

Tool steels are ferrous alloys used for shaping and cutting other materials. They contain more alloys than HSLA steels and, in most instances, combine hardenability with the presence of hard, second-phase **carbides**. Careful preparation and control in processing these alloys is absolutely necessary, so most tool steels are melted in electric arc furnaces and many remelted in a vacuum to ensure the cleanliness required. Small quantities are the norm and total quality management is essential. Most tool steels are available only as bar, rod, or large forgings.

There are many grades of tool steels, but they can be classified in four main categories:

> cold work
>
> hot work
>
> shock resistant
>
> high speed

These classifications reflect their applications. For example, all the **cold work tool steels** are hardenable and this category can be differentiated further into water hardenable, oil hardenable, and air hardenable tool steels.

Water hardenable tool steels are hypereutectoid steels containing up to 1.5% C plus minor amounts of carbide formers Cr, Mo, W, and V. These alloys yield high surface hardness with a tough interior and can provide sharp, wear-resistant cutting edges. Oil hardenable tool steels have increased Mn, Cr, and W, which not only shifts the nose of the isothermal transformation curve to longer times, making the steel more hardenable, but also forms some carbides. These alloys are used for blanking and drawing dies. Air hardenable tool steels contain higher amounts of Cr, Mo, and V and, in some cases, carbon itself. Figure 7.8 illustrates the microstructure of D2, an air hardenable cold work tool steel.

Figure 7.8
Microstructure of D2 cold work tool steel. Globular white phase is the carbide. (1000×) (Courtesy of Latrobe Steel Company.)

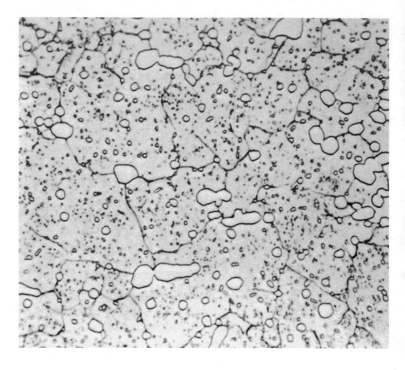

Hot work tool steels are used for handling red-hot ingots and billets during hot-working deformation processes and for molds used in injection molding of plastics. These steels have medium carbon content and contain moderate amounts of W, Cr, V, and Mo. The carbides reduce softening at elevated temperatures. Chromium types contain about 5% Cr and may contain lesser amounts of V, Mo, and W. The desired microstructure for H13 appears in Figure 7.9. Tungsten types contain 9–18% W and Cr; molybdenum types contain about 5% each of Mo, Cr, and W.

Shock-resistant tool steels were developed for the toughness necessary in shock applications, such as cold chisels and crowbars. They have compositions similar to cold work tool steels, but have lower carbon content, which produces through-hardened parts with hardness somewhat less than Rc 60.

Perhaps the best-known tool steels are the **high-speed tool steels**, which are used for machining other metals at high speeds. These alloys contain high carbon plus the largest concentrations of Cr, V, Mo, and W. The result is a hard matrix containing a dispersion of numerous carbides. These highly alloyed steels resist softening at elevated temperatures, thus maintaining a good cutting edge during machining, and are durable because they resist wear during machining. There are two main types of high-speed steels, identified by *M* for molybdenum and *T* for tungsten, two elements that are the most important for high-temperature strength. Tungsten types contain more than 12% W; molybdenum types replace tungsten with the amount of Mo equal to about half the amount of W being replaced. All the high-speed steels can be hardened to Rc 62–67. Their microstructure is a fine dispersion of the carbides in a tempered martensite matrix, such as that shown for M7 high-speed tool steel in Figure 7.10. Hardness is maintained at temperatures up to 1000°F.

Figure 7.9
Microstructure of H13 hot work tool steel (1000×) (Courtesy of Latrobe Steel Company.)

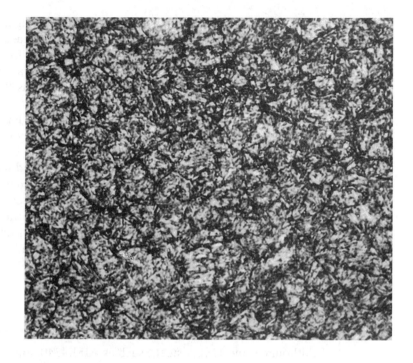

Figure 7.10
Microstructure of M7 high-speed tool steel (1000×) (Courtesy of Latrobe Steel Company.)

      Criteria that are usually applied when selecting a tool steel are based on hardening and use. Hardening characteristics include ease of hardening, distortion and cracking tendencies, plus decarburization problems for high-carbon alloys. Use characteristics include wear resistance, high-temperature strength, toughness, and machinability. These characteristics are the same as those we have been examining throughout this text — properties and how to control them. Hardening characteristics are best for air hardenable alloys that have the least distortion and best hardenability. However, these are cold work steels, so they have limited use. On the other hand, high-speed steels have the worst hardening characteristics and poor toughness, but their high-temperature strength associated with fine carbide dispersion in tempered martensite provides their important use characteristics.

      We optimize both the hardenability and use characteristics through chemical composition, balancing the carbon and carbide formers and the appropriate heat treatment in order to achieve the desired microstructure. The chemical compositions of some selected tool steels are summarized in Table 7.4.

7.5 *Maraging Steels*

Although some HSLA steels, such as AISI 4340, and some tool steels, such as H11, can be heat-treated to ultrahigh strength levels, they have limited ductility and, therefore, poor toughness. **Maraging steels** are Fe-Ni alloys that have ultrahigh strength achieved by aging of carbon-free martensite. Table 7.5 lists the nominal composition of some of these alloys.

      Early studies demonstrated the superior fracture toughness of maraging steels, which are compared to HSLA and H11 tool steel in Figure 7.11. These steels are

Table 7.4
Chemical composition of selected tool steels

Type	Alloy designation	C (%)	Cr (%)	V (%)	W (%)	Mo (%)
Cold work	W4	0.6–1.4	0.25	—	—	—
	O1	0.9	0.5	—	.5	— (1 Mn)
	D2 (air)	1.5	12.0	1.0	—	1.0
	A2	1.0	5.0	—	—	1.0
	A7	2.25	5.25	4.5	—	1.0
Hot work	H13	0.35	5.0	1.0	—	1.5
	H42	0.6	4.0	2.0	6.0	5.0
	H21	0.35	2.0	—	9.0	—
Shock resistant	S1	0.5	1.5	—	2.5	—
High-speed	M7	1.0	4.0	2.0	1.75	8.75
	T1	0.7	4.0	1.0	18.0	—

Table 7.5
Nominal composition of selected maraging steels

Alloy designation	Nominal composition (wt %)				Yield strength (psi)
	Ni	**Co**	**Mo**	**Ti**	
200 alloy (ASTM grade A)	18	8	3	0.2	200,000
250 alloy (ASTM grade B)	18	8	5	0.4	250,000
250 alloy (Co-free)	18	—	3	1.3	250,000
300 alloy (ASTM grade C)	18	9	5	0.7	300,000
350 alloy	18	12	4	1.6	350,000
400 alloy	13	15	10	0.2	400,000

used for military applications, aerospace rocket motor cases, pressure hulls for deep-sea exploration vessels, and extensive production tooling for manufacturing.

Heat treatment of maraging steels begins with solutionizing at temperatures of about 1800°F (1000°C). If the alloys are impure, Ti(C, N) precipitates and embrittles the steel. During cooling, the austenite decomposes, forming a precipitate of $Fe_2(Mo, Ti)$ in much the same way that austenite decomposes in mild steel to form pearlite. The martensite that is formed, however, is a lath type martensite characterized by high dislocation densities. Maximum strength is developed by aging the martensite at temperatures of about 900°F. Strength is derived from the fine size of the precipitates, about 0.005 micron. These precipitates are intermetallics of (Fe, Ni) and (Ti, Mo); the role of Co apparently is synergistic, but Co-free alloys with yield strengths up to 300,000 psi have been developed by increasing the concentrations of both Ti and Mo.

There are many advantages to the maraging steels besides high strength and superior fracture toughness. They retain their strength to high temperatures, are

Figure 7.11
Effect of strength level on fracture toughness of high-strength steels

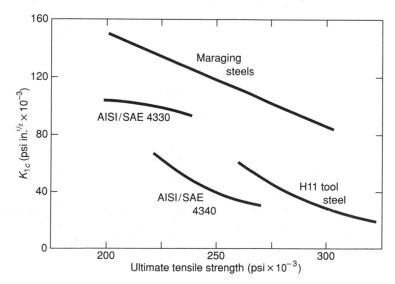

machinable before hardening, are not subject to quench cracking, are weldable even in the aged condition, and can be cold-worked as well as hot-worked. They also are economical because of savings in machining and heat treating.

These advantages have led to increased applications only because of innovations in melt practice and thermomechanical processing (which we will learn more about in Chapter 9). Impurities have been reduced by improved melting techniques, involving not only vacuum induction, vacuum arc melting, and electron beam melting, but also computer programmed carbon/oxygen boils. Consequently, impurities such as C, O, N, and S have been lowered to less than 30 ppm total. Thermomechanical processes such as **ausforming** and **marforming**, that is, deforming the austenite or martensite, or combinations thereof have been used to reduce the austenite penultimate grain size, thus generating fine martensite with high dislocation content while dispersing any nonmetallic and intermetallic phases that may have formed. Figure 7.12 demonstrates how ausforming carried out during cooling has been extended to low temperatures by quenching to eliminate precipitation of intermetallics and further ausforming before cooling to form the martensite. Such processing has developed strengths of 500,000 psi with some ductility.

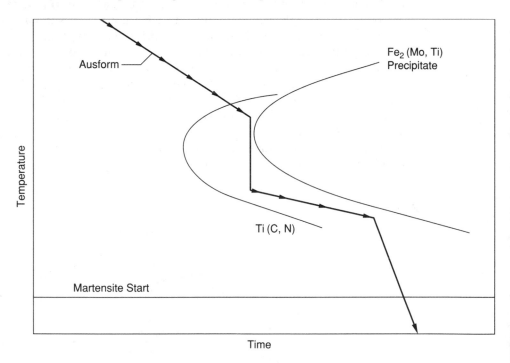

Figure 7.12
Ausforming of maraging steel to disperse Ti(C, N) particles and refine austenite grain size, but retain Mo and Ti in solution for subsequent maraging
(From R. F. Decker and S. Floreen, in *Maraging Steels, Recent Developments and Applications*, edited by R. K. Wilson, The Metallurgical Society, AIME, 1988.)

Case Study 7.3

Maraging Steels and the Olympics

Preparation behind the scenes for any athletic event is not unlike manufacturing and management relationships. When the event is the Olympic Games, however, the stakes are much higher and the preparation takes on a different pitch. Technology and materials have played a dramatic role in improving sports equipment, leading to many new world and Olympic records in such diverse events as springboard diving, crossbow archery, pole vaulting, and others.

Safety was the driving force for a materials change in fencing equipment for the most recent Olympics. The most dangerous situations in competition arise because the foils or épées can break during the heat of competition. The broken sections are stiffer than the unbroken foil and the fractured ends are sharp. Injuries have been caused when the sharp ends punctured fencers' protective equipment.

Traditionally, blades have been made from carbon steels. In a world-class meet, as many as one hundred blades have snapped during competition. Concerned sportsmen have since learned about the superior toughness of maraging steels, and now épées and foils are made only from these alloys. In the 1991 world championship competition, only three maraging steel blades broke. (*Note:* In volume two, you'll learn that the personal protective equipment for fencers has also resulted from advances in materials.)

7.6 Ferrous Magnetic Alloys

As we learned in Chapter 1, there are three types of magnetism — ferromagnetism, paramagnetism, and diamagnetism — but only ferromagnetism is important from our viewpoint of the interrelations of selection, processing, and properties. Within the ferromagnetic classification, there are two important types of materials, so-called hard and soft. The hard magnetic alloys are made into **permanent magnets** and the **soft magnetic alloys** are used in electrical machinery and power generation equipment. There are many ferromagnetic materials of both types that are not ferrous based, such as the ceramic ferrites, nickel alloys that used for magnetic shielding, and even Fe-Ni alloys that contain as much iron as nickel. We consider these to be nonferrous because they are austenitic where Fe atoms behave very differently than in bcc crystal structures.

7.6.1   Soft Magnetic Electrical Steels

Electrical steels are rolled sheet products available in thicknesses from 0.010 in. to 0.027 in. They are particularly fascinating because most do not contain carbon, and processing plays a most important role in achieving optimum properties. Even carbon steels, referred to as lamination steels, differ from ordinary carbon steels because they contain Mn and P to increase electrical resistivity and must be processed in a manner to promote grain growth in annealing. These alloys are used in small motors (<1 horsepower).

Electrical steels that contain silicon in concentrations from 0.5% to 3.25% and aluminum additions up to about 0.5% are divided into two categories, nonoriented and oriented. Nonoriented steels are used in a wide variety of equipment, such as generators, motors, lamp ballasts, relays, and transformers. Most oriented steels are used in transformers for electric utility systems. In these applications, energy lost as heat when the material is magnetized is the most important criterion. These losses are called **core losses** and have been estimated to account for 4–5% of the total of all electricity generated. Core loss can be divided into two categories, hysteresis losses and eddy current losses. Silicon is an important component of soft electrical steels because it increases resistivity and lowers coercive force, thus effectively reducing both types of losses.

The lowest core losses for nonoriented steels are achieved by controlled thermomechanical processing (see Chapter 9) to remove all the carbon and to promote grain growth in the annealed sheet. Oriented grades have lower losses because of their **anisotropy**, taking advantage of the easier magnetization in certain crystal directions. These alloys, however, require meticulous control of thermomechanical processing to develop texture (again, see Chapter 9), but they also contain small additions of AlN to inhibit grain growth and promote secondary recrystallization. This treatment is necessary to reduce the core losses below those of nonoriented steels.

Another discovery that has further reduced the core losses for oriented steels is the application of stress coatings. Although all electrical steels must have an insulating coating on their surfaces so they can behave magnetically as individual sheets, it is possible to have these coatings place the surface in tension. Figure 7.13 demonstrates these reduced losses as a function of the induced field. We believe that the lower core loss for the stress-coated high-perm alloy is associated with reduced energy that permits magnetic domain movement during the magnetization process. This is analogous to dislocation motion easing the deformation process, which we will learn about in Chapter 9.

7.6.2   Permanent Magnetic Alloys

A permanent magnetic alloy has high remanence, usually reported as the maximum product of magnetic fields B and H, $(BH)_{max}$, and high coercive force, H_c. The num-

Figure 7.13
Superior performance of stress-coated high-perm H2 alloy compared to standard grain-oriented alloys (From F. E. Werner, "Electrical Steels: 1970–1990," in *Energy Efficient Electrical Steels*, edited by A. R. Marder and E. T. Stephenson, The Metallurgical Society, AIME, 1980.)

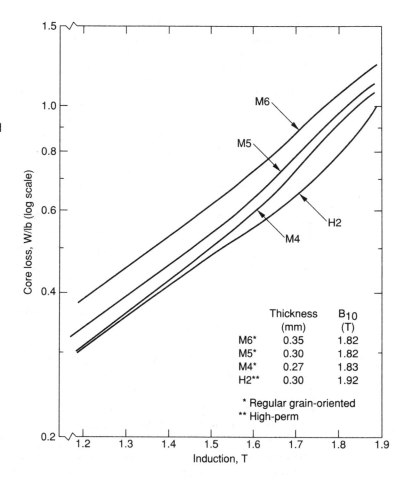

	Thickness (mm)	B_{10} (T)
M6*	0.35	1.82
M5*	0.30	1.82
M4*	0.27	1.83
H2**	0.30	1.92

* Regular grain-oriented
** High-perm

ber of ways we use these alloys is not only large, but extremely diverse, as shown in Table 7.6.

The earliest permanent magnetic alloys, hardenable carbon steels, were followed by Fe-6% W and Fe-Co alloys that contained up to about 35% Co; all of these were hardenable and could be hot-worked for shaping. They were easily demagnetized, however, and lost favor when **alnico** alloys (containing Al, Ni, and Co) were first developed in 1931. These alloys are brittle and are available only as cast shapes or as made by powder metallurgical techniques (covered in volume two). Although the first alnico had no cobalt, subsequent alloys contain up to 40% Co. Additions of Cu and Nb increase coercivity and titanium has proved helpful in preventing precipitation of fcc-δ at high temperatures. The magnetic properties of the original alloys were sensitive to cooling rates and subsequent heat-treating associated with the δ-precipitation and its effect on coarsening microstructure when it transforms to the bcc phase at lower temperatures.

Table 7.6
Some applications of permanent magnetic alloys

Magnetic characteristic	Application
Electromechanical forces	Compass (torque) Holding devices (attraction) Bearings, levitation (repulsion) Brakes and leveling devices (electromagnet/magnet combination) Metal separators
Electromagnetic forces	Loudspeakers Hysteresis motors Some telephone receivers
Electromagnetic induction	Microphones, sensing devices Eddy current testers for welds
Forces on moving electrons	Focusing of electron microscopes

Today, most alnico alloys containing more than 20% Co are cooled quickly to avoid δ-precipitation, which occurs at about 1100°C, then a magnetic field is applied through the temperature range of 850–750°C in the preferred direction for the final part. Decomposition into two bcc phases begins in this temperature range and the alloy is below its Curie temperature because of the cobalt content. Figure 7.14 illustrates the extremely fine two-phase microstructure parallel and perpendicular to the applied field that is developed. The two bcc phases have nearly the same crystal structure but different compositions that alter their magnetic characteristics, producing very high coercivity while maintaining high $(BH)_{max}$. Although all alnicos are presently treated in this manner, they are referred to as anisotropic rather than isotropic where no magnetic field is applied during cooling.

Compositions and magnetic properties of permanent magnetic alloys are summarized in Table 7.7.

Summary

Ferrous alloys include all metal alloys whose properties are based on the major constituent, iron. The most widely used ferrous alloys are mild steels that are $Fe-Fe_3C$ alloys, used extensively for structural purposes. High-strength low-alloy steels provide higher strength through heat treatment. Cast irons are higher carbon alloys that are brittle but can be cast economically because of their lower melting temperature. These materials have the microstructure of steel but contain free graphite as well. Adding 12% or more chromium to iron imparts corrosion resistance. Stainless steels utilize the addition of Cr, but they can also be strengthened by alloying or heat treat-

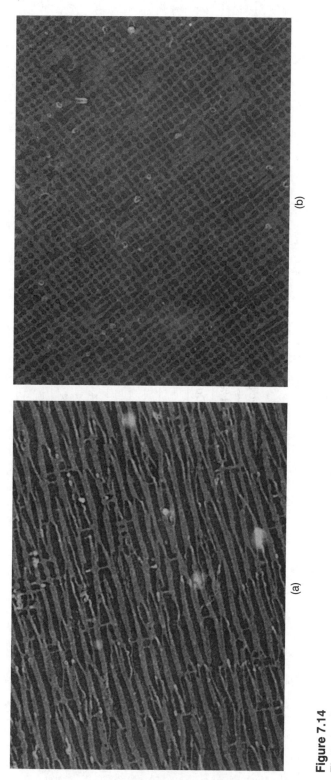

Figure 7.14

Electron micrographs of the two-phase microstructure of alnico held isothermally in a magnetic field for 9 min at 800°C: (a) parallel to field direction, (b) perpendicular to field direction (50,000×) (From K. J. de Vos, The Relationship Between Microstructure and Magnetic Properties of Alnico Alloys, thesis, Eindhoven.)

(a)

(b)

Table 7.7

Composition and magnetic properties of selected permanent magnetic alloys

Material	Nominal composition (wt %)										$(BH)_{max}$ (MGOe)	H_c (Oe)
	C	Cr	W	Al	Ni	Co	Cu	Nb	Ti	Fe		
1% C steel	1.0	—	—	—	—	—	—	—	—	Bal.	0.20	50
6% W	0.7	0.9	6	—	—	0.5	—	—	—	Bal.	0.30	65
6% Cr	1.0	6	—	—	—	—	—	—	—	Bal.	0.30	65
35% Co	0.8	6	4	—	—	35	—	—	—	Bal.	0.95	250
Alnico, A1*	—	—	—	10	16	0	—	—	—	Bal.	1.25	470
Alnico, A2*	—	—	—	9	16	26	4	1	—	Bal.	5.5	650
Alnico, A3*	—	—	—	9	16	26	4	3	1	Bal.	4.5	780
Alnico, A5*	—	—	—	8	16	40	4	2	8	Bal.	5.8	2000

* Anisotropic

ment for many applications. Tool steels have been developed for shaping and cutting other metal alloys. They must be hardened to Rc 60s and must maintain hardness to elevated temperatures in certain applications. These features are achieved with microstructures consisting of tempered martensite that contains dispersed carbides. Ultrahigh strength can be achieved for maraging steels, which form a carbon-free martensite that can be aged to precipitate extremely small intermetallics. These alloys are also used extensively for tooling applications. Many ferrous alloys are used because of their magnetic properties, either for soft magnetic applications as electrical steels or as permanent magnets. Electrical steels are Fe-Si alloys that are carbon-free. They are used in alternating current devices such as transformers, so low core loss is important for energy conservation. This is achieved by control of composition and thermomechanical processing. Permanent magnetic alloys are predominantly brittle and available only as castings. The most popular alloys are alnicos, which must be heat-treated properly, at times in the presence of a magnetic field, in order to achieve the desired properties.

Terms to Remember

AISI/SAE	gray cast iron
alnico	high-speed tool steel
anisotropy	hot shortness
API	hot work tool steel
ASTM	malleable cast iron
ausforming	maraging steel
carbides	mild steel
cast iron	music wire
cold work tool steel	permanent magnet
core loss	plain carbon steel
ductile cast iron	shock-resistant tool steel
ductile-to-brittle transition	soft magnetic alloys
electrical steel	temper carbon
graphite flakes	

Problems

1. List two applications for each of the following:
   a. mild stecl
   b. HSLA steel
   c. austenitic stainless

    d. high-speed steel

    e. nonoriented electrical steel

    f. maraging steel

    g. alnico

    h. gray cast iron

2. Explain what effect graphite flakes have on the properties of gray cast iron.

3. Describe the heat treatment for malleable cast iron.

4. Explain the differences among the types of tool steels.

5. Compare the performance of a carbon steel drill bit with that of a drill bit made of high-speed tool steel.

6. Compare the precipitation that takes place in maraging steels to that described for aluminum alloys in Chapter 6, Section 6.5.

7. Explain what core losses are.

8. Explain the difference between oriented and nonoriented electrical steels from the viewpoint of

    a. applications

    b. processing

9. Explain why oriented electrical steels are glass coated.

10. Draw a TTT diagram for alnico, demonstrating the proper cooling from 1200°C and the decomposition phases that must be avoided. (*Hint:* Use Figure 7.12 as a guide.)

8

Nonferrous Metals and Alloys

Nonferrous is an all-inclusive term that encompasses all metals except iron and all alloys that do not have iron as the major element. Applications for nonferrous alloys are more diverse than for ferrous alloys because the characteristics of the major elements are more diverse. For example, strength of ferrous alloys is an overwhelming consideration for most applications. We will examine how strength can be improved in nonferrous alloys, but their strength, with few exceptions, is inferior to that of ferrous alloys and is a major concern only for superalloys used at elevated temperatures (e.g., turbines for jet engines) and beryllium-copper, used for its spring characteristics. Strength combined with low density or strength combined with corrosion resistance, however, is important for many nonferrous alloys such as titanium- and aluminum-base alloys. We will find electrical conductivity, corrosion resistance, color, melting temperature, and manufacturability to be major concerns for nonferrous applications. Magnetic properties will also be important, but for different reasons than for electrical steels and permanent magnets.

8.1 Copper and Copper Alloys

Copper is an fcc metal that is an important engineering material in its pure form as well as a base for commercial alloys. Applications for pure copper are based on excellent electrical and thermal conductivity, ease of fabrication in many forms and ease of joining, plus good corrosion resistance and reasonable strength. As a base for alloying, copper is best known for single-phase solid-solution alloying with zinc to form brass and with tin to form bronze. Precipitation hardenable alloys such as copper-beryllium alloys with strengths greater than 200,000 psi are also possible, as are bearing alloys for wear applications.

The classification system for copper and copper alloys most widely accepted is that of the Copper Development Association. This system is outlined in Table 8.1.

8.1.1 Pure Copper

Pure copper is used for its excellent electrical conductivity as wire or bus bars in electrical applications and as printed wiring in electronic applications. It is also used for its thermal conductivity in pipe for hot water heating systems and for protection of superconducting filaments should they go normal (i.e., lose superconductivity) while carrying current; otherwise, the filaments would burn out from the sudden

Table 8.1
Classification of copper and its alloys

Wrought alloys	Cast alloys	Number
Coppers (>99.3% Cu) and high-Cu alloys (96–99.3% Cu)		C1xxxx
Cu-Zn alloys (brasses)		C2xxxx
Cu-Zn-Pb alloys (leaded brasses)		C3xxxx
Cu-Zn-Sn alloys (Sn brasses)		C4xxxx
Cu-Sn alloys (phosphor bronzes)		C5xxxx
Cu-Al alloys (Al bronzes), Cu-Si alloys (silicon bronzes), and miscellaneous copper-zinc alloys		C6xxxx
Cu-Ni (Cu nickels) and Cu-Ni-Zn alloys (Ni silvers)		C7xxxx
	Cast coppers, high-Cu alloys, brasses, leaded brasses, Mn bronzes, and Si bronzes	C8xxxx
	Cast Sn bronzes, leaded bronzes, Ni-Sn bronzes, Cu nickels, Ni silvers, Al bronzes, special alloys	C9xxxx

resistance heating. Refined copper, known as electrolytic tough pitch (ETP) copper, contains as much as 0.2% oxygen. ETP copper is used as a base for alloying, but cannot be used where highest conductivity is necessary. In those cases, oxygen-free high-conductivity (OFHC) copper is used.

OFHC copper is ductile but not strong, having a tensile strength of about 32,000 psi and elongation of about 55%. It can be strengthened by cold work to tensile strengths of about 65,000 psi, but elongation is reduced to as low as 5%. Of course, intermediate anneals (see Chapter 10) can be employed to control the properties for processing as well as for final applications, making fabrication of copper and copper alloy parts an easy task.

8.1.2 Brass Alloys

Brass is the alloy of copper and zinc. We learned in Chapter 4 that the solubility of zinc in brass is about 40%. The strength in the annealed condition is increased from the copper values to above 50,000 psi while maintaining elongation values of about 45%. Cold-worked alloys have strengths up to 80,000 psi, with about 5% elongation. The best-known brass is the 30% Zn alloy, known as **cartridge brass**, used in ammunition but also for musical instruments and hardware. **Yellow brass**, containing 35% Zn, is used for springs and screws. Machinability is improved by up to 4% Pb additions for automatic screw manufacture. **Muntz metal** contains 40% Zn, the highest amount in alloys used structurally; it is used in architectural applications and for condenser tubes.

8.1.3 Bronze Alloys

Bronze is the oldest alloy known to man. Used in the cast form by primitive man during the Bronze Age, today it is still used for decorative castings and for gears, pipe fittings, and pump parts on ships because of excellent corrosion resistance to seawater. Few wrought bronze alloys are used, however, because of the work-hardening characteristics and the high cost of tin.

Most bronzes are copper-tin alloys containing up to 12% Sn, the limit of solid solubility. Other bronzes contain no tin, but derive their name from the characteristic bronze color. Therefore we have aluminum bronzes that contain aluminum plus smaller amounts of iron or silicon and silicon bronzes that contain up to 4% silicon.

Copper-tin bronze castings are used for bearings, gears, and fittings because they have good wear resistance, strength, and corrosion resistance. When used as bearings, lead is added. The most common wrought bronze is phosphor bronze, which is single-phase Cu-Sn with 0.15% P added for deoxidation. These alloys can be work-hardened to about 110,000 psi, but elongation is reduced to about 3%.

8.1.4 Copper-Nickel Alloys

We learned in Chapter 3 that copper and nickel form the only alloys with complete solid solubility. These alloys all have high thermal conductivity, excellent corrosion resistance, and high strength. Copper-base alloys are the cupronickels and nickel silvers; the latter contain zinc as well as nickel additions to copper. **Cupronickels** are typically used for heat exchange, whereas nickel silvers are used for slide fasteners and camera parts because of their cosmetic appearance (implied by the name).

Nickel-base alloys are known as monels and have higher strengths than cupronickels combined with excellent corrosion resistance. Their main utility is in diverse marine applications.

Case Study 8.1

Selecting the Right Contact

Modern electronic switches are simple devices that are controlled by software, in many instances for uses in sophisticated electronic equipment. A switch by definition provides intermittent contact between two components, at which time current flows through the contact from one component to the other. Switches such as the typical one shown in Figure 8.1 are susceptible to wear because of repeated contact and to oxidation; both increase the contact resistance and decrease performance.

When a multiwire switch for an electronic instrument was being designed at Tekworth, Inc., with manufacturability in mind, reliability required careful selection of the alloy for the ground plane. Copper was considered, but its ductility and corrosion characteristics were deterrents (despite the reasonable corrosion resistance of copper). In surveying metal alloys available from several vendors, the manufacturing supervisor at Tekworth combined the vendor data, plotting the contact resistance of clean and exposed samples, as shown in Figure 8.2. The supervisor's efforts made selection much easier. Alloy C63800, which is an aluminum bronze, was selected over alloy C69000, a miscellaneous brass alloy containing aluminum, after it was determined that the mechanical properties were similar for these two alloys.

A preliminary test for accelerated exposure of prototype switches nevertheless was conducted using copper and the two alloys. Although both alloys were superior to copper, neither was found to be acceptable. The contact surfaces were examined in a scanning electron microscope, which revealed severe contamination on the surfaces of all samples — contamination such as oil and skin tissue related to worker handling during assembly. After establishing clean-room assembly conditions, C63800 alloy was confirmed as the preferable choice.

Figure 8.1
Typical contact geometry:
(a) open circuit, (b) closed
circuit

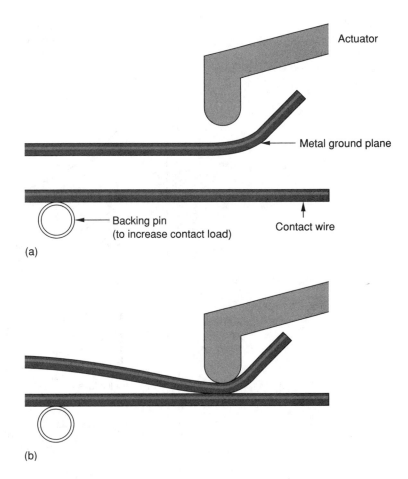

Actuator

Metal ground plane

Backing pin
(to increase contact load)

Contact wire

(a)

(b)

8.1.5 *Precipitation-Hardened Copper Alloys*

We learned in Chapter 6 how precipitation from a supersaturated solid solution could enhance the strength of binary alloys such as aluminum-rich Al-Cu alloys. There are two binary copper-base alloys for which we also take advantage of this strengthening method. The best known is **beryllium-copper**, which contains 2% Be and can achieve tensile strengths of about 180,000 psi with 5% elongation. The alloy is nonmagnetic, nonsparking, corrosion resistant, and has high conductivity; it is useful for springs, electrical contacts such as fuse clips, bellows, and diaphragms. Caution is recommended, however, because the toxicity of beryllium has limited any interest in these alloys in recent years.

Other age hardenable copper-base alloys are Cu-2% Ni-0.6% Si, Cu-0.8% Cr, and Cu-0.2% Zr alloys. These alloys can be strengthened to tensile strengths of about 70,000 psi by combining cold work with the aging heat treatments.

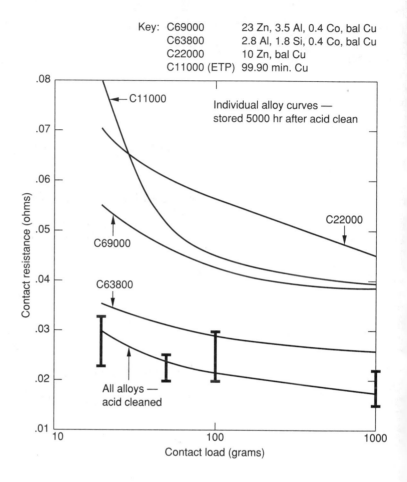

Figure 8.2

Influence of contact load on contact resistance of clean and oxidized contact materials

Key: C69000   23 Zn, 3.5 Al, 0.4 Co, bal Cu
      C63800   2.8 Al, 1.8 Si, 0.4 Co, bal Cu
      C22000   10 Zn, bal Cu
      C11000 (ETP)   99.90 min. Cu

8.2   *Aluminum and Aluminum Alloys*

Aluminum and aluminum alloys are the most widely used nonferrous metals, with applications in construction, transportation, containers, electrical equipment, household items such as cookware and foil, and mechanical equipment. These applications are the result of the unique properties of aluminum, particularly excellent electrical and thermal conductivity, light density, and corrosion resistance. In addition, aluminum ore is abundant and aluminum is readily recycled; it is easily deformed and cast into many diverse shapes.

Perhaps the only limiting features of aluminum are the relatively low melting point and low modulus of elasticity. Both density and modulus of elasticity are about one third the value of steel. Thus aluminum will deflect more than steel under the same loading and we need to design structural parts with this in mind. However, the specific strength, that is, the strength-to-weight ratio, for precipitation hardenable aluminum alloys is superior to that of steel. We take advantage of this in aircraft,

Table 8.2
Classification of aluminum
and aluminum alloys

Wrought alloys	Cast alloys	Number
Al, 99.00% min.		1xxx
Major alloy addition		
Cu		2xxx
Mn		3xxx
Si		4xxx
Mg		5xxx
Mg and Si		6xxx
Zn		7xxx
other elements than above		8xxx
	Major alloy addition	
	Cu	2xx
	Si with Cu and/or Mg	3xx
	Si	4xx
	Mg	5xx
	Zn	7xx
	Sn	8xx
	other elements	9xx

where aluminum alloys find extensive use because of the specific strength combined with inexpensive cost. Aluminum alloys do have some drawbacks, though, such as poor wear resistance, difficulty in metallurgical joining, and poor fatigue resistance (as we saw in Chapter 1). Stress concentration due to porosity is a particular problem for cast aluminum alloys.

8.2.1 Alloy Designation, Composition, and Typical Properties

The Aluminum Association uses separate numbering systems for classifying wrought and cast aluminum, as Table 8.2 indicates.

Pure aluminum is not included as a cast metal in Table 8.2. It is not used in the cast form because of high shrinkage and susceptibility to hot cracking. Table 8.3 gives some typical compositions, applications, and properties for popular commercial alloys.

8.2.2 Heat Treatment of Aluminum Alloys

Most but not all of the higher strength aluminum alloys are age hardenable. This precipitation hardening is related to their binary phase diagrams, which have eutectic reactions for aluminum-rich compositions. The Al-Cu eutectic is

$$\ell(33.2\% \text{ Cu}) \rightarrow \kappa(5.65\% \text{ Cu}) + \theta(47.5\% \text{ Cu}), \ T = 548°C$$

Table 8.3
Compositions, applications, and properties for commercial aluminum alloys

Alloy type	Alloy number	Composition (wt %)					UTS (ksi)*	Percent elongation*	Applications
		Cu	Si	Mg	Mn	Other			
Wrought	1100	0.12	—	—	—	—	13(O)	13(O)	Foil, conductor wire
	2024	4.4	—	1.5	0.6	—	27(O)	10(T6)	Aircraft parts
	3004	—	0.18	1.0	1.1	0.4 Fe	26(O)	22.5(O)	Recycled cans
	5052	—	—	2.5	—	0.25 Cr	28(O)	8(H)	Rivets, fuel lines
	6061	4.4	—	1.5	0.6	—	18(O), 45(T6)	12(T6)	Treadplate, railings
	7075	1.6	—	2.5	—	5.6 Zn, 0.25 Cr	38(O), 83(T6)	11(T6)	Aircraft parts
Cast	295	4.5	1.0	—	—	1.0 Fe	32(T6)	5(T6)	Flywheels
	356	—	7.0	0.3	—	—	30(T6)	3(T6)	Automotive parts
	380**	3.5	8.5	—	—	3.0 Zn, 2.0 Fe	46	3.5(F)	Housings
	413**	—	12.0	—	—	2.0 Fe	43	2.5(F)	Intricate castings
	518**	1.0	—	8.0	—	1.8 Fe	45	6.5(F)	Marine ornaments
	771	—	—	—	—	7.0 Zn	42(T6)	5(T6)	Computer parts
	850	1.0	—	—	—	6.3 Sn, 1.0 Ni	23(T5)	10(T5)	Bearings

*See Chapter 6 for temper terminology and designations (e.g., T6).
**Die-cast alloys

the Al-Mg eutectic is

$$\ell(35.0\% \text{ Mg}) \rightarrow \alpha(14.9\% \text{ Mg}) + \beta(35.5\% \text{ Mg}),\ T = 451°C$$

and the Al-Si eutectic is

$$\ell(12.6\% \text{ Si}) \rightarrow \alpha(1.65\% \text{ Si}) + \text{Si}(99.83\% \text{ Si}),\ T = 577°C$$

Wrought alloys in Table 8.3 containing these elements — 2024, 5052, 6061, 7075 — are all age hardenable. All of the cast alloys in Table 8.3 are age hardenable. Commercial heat treatments are usually long duration, low-temperature aging (see Figure 6.17) to promote uniform properties throughout the material. Because diffusion does occur at low temperatures in these alloys, the influence of temperature variations in service on dimensional stability and properties must be considered.

Heat treatable alloys have reduced corrosion resistance because of their two-phase microstructure. When high strength and corrosion resistance are needed, alloy 1100 is roll-bonded to the surface of the age hardenable alloy during processing to form Alclad, a product that provides both corrosion resistance and strength.

8.2.3 Non–Heat Treatable Aluminum Alloys

Other alloys are strengthened by work hardening, which we will study in Chapter 9. Notable among these alloys is 3004. Manganese was found to be one of the few solid-solution strengthening alloy additions that did not reduce the corrosion resistance of aluminum. The composition of 3004 aluminum alloy, nominally, 1.1% Mn, 1.0% Mg, 0.4% Fe, and 0.18% Si, was developed for beverage cans as a compromise of formability, strength, and recycling potential. This alloy does undergo eutectic decomposition of the liquid phase, but there is no solubility to permit age hardening. The eutectic microstructure is broken up during hot and cold working, leaving a matrix containing about 2.5% of fine dispersoids, as shown in Figure 8.3.

The composition of alloy 3004 gives a compromise of strength and formability for beverage cans and recycling ability. In the hardened condition, sheet is drawn into a cup, then redrawn and ironed through a series of steps, finishing with the dome, neck, and flange at a rate of about 200 cans per minute, with a failure rate of less than 1 in 10,000.

Case Study 8.2

Failure of 356 Aluminum Alloy Flywheel

A new chain saw ordered by a large manufacturing firm was inspected and tested by their maintenance department upon receipt and was returned to the distributor because the lubrication mechanism did not work. Back at the distributor, a technician removed the

Figure 8.3
Dispersoids in cold-rolled 3004 aluminum alloy sheet (From A. K. Vasudevan and R. D. Doherty (eds.), *Treatise of Materials Science and Technology*, Vol. 31: *Aluminum Alloys — Contemporary Research and Applications*, Academic Press, 1989.)

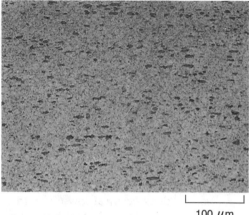

100 μm

arbor and chain, placed the chain saw in a vise, and started the motor to check the lubrication. At about three-quarter throttle, the saw literally self-destructed by exploding. Fortunately no one was injured.

The parts were collected and sent to the manufacturer to determine why the motor failed. It was found that the flywheel fractured, as shown in Figure 8.4. The flywheel was designed to withstand the high centrifugal forces, however, so a metallurgical investigation was begun. The alloy was identified as 356 aluminum, typically used for such parts. Microscopic examination showed the failure probably initiated at an apex of the hexagonal hub that fits onto the crankshaft assembly. The fracture was brittle and dark streaks were observed on the fracture surface at low magnifications. Scanning electron microscopy revealed microcracking in the brittle fracture area and high Fe concentration. These microcracks are shown in Figure 8.5 and the corresponding X-ray map for Fe is shown in Figure 8.6.

The aluminum die-casting foundry was advised of these findings and measures were taken to dissolve the iron more efficiently in future production.

8.3   *Nickel and Nickel Alloys*

Nickel is important both as a pure metal and as a base for commercial alloys with diverse applications. Pure nickel is fcc and has good electrical conductivity in addition to good strength and corrosion resistance. Many applications are found for the pure metal in plating, electrical contacts, electronic components, and food processing equipment.

Figure 8.4
Fractured parts of aluminum alloy flywheel

Nickel alloys include the monels, which are Ni-rich solid solutions containing Cu. These are particularly useful for marine applications that take advantage of the good corrosion resistance to seawater. The addition of small amounts of Al and Ti to monel can provide precipitation hardening by formation of Ni_3Al or Ni_3Ti. Many ultra-high-strength alloys, called **superalloys** because of their strength retention at high temperatures, have also been developed. These superalloys find numerous high-temperature applications, for example, as jet engine parts. The magnetic characteristics of fcc Ni-Fe alloys have led to numerous applications in electronics, magnetics, and in joining glass to metal. The properties of interest range from the highest magnetic permeability, used in pulse transformers and in magnetic shielding, to controlled low thermal expansion necessary for thermally matched glass-to-metal seals.

Figure 8.5
SEM micrograph of the brittle fracture surface at the apex of the hexagonal hub of the flywheel. The arrows indicate microcracks. (150×)

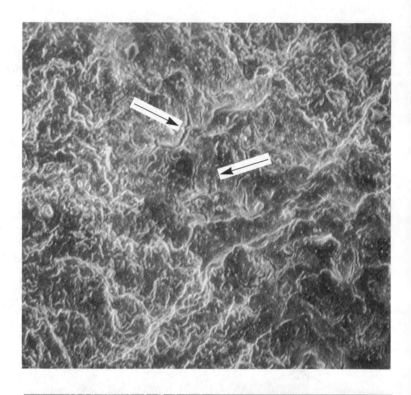

Figure 8.6
SEM X-ray map of Fe in the fracture surface of Figure 8.5 (150×)

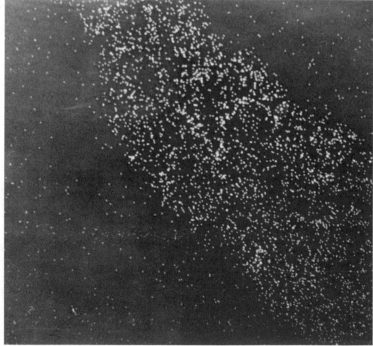

8.3.1   High-Temperature Alloys

In addition to hot work tool steels, which we studied in Chapter 7, there are many high-temperature applications for alloys, for example, jet engines and rocket engines in aeronautics and aerospace industries. High strength at the high temperatures is needed, of course, but creep strength, which is resistance to deformation at the high-temperature stress, and oxidation resistance are also imperative. Although there are some ferrous and some cobalt-base high-temperature alloys, the most common ones are nickel-base.

Wrought nickel-base superalloys are double-melted in vacuum induction furnaces, then remelted by vacuum arc to form an ingot. These alloys can also be cast to shape, using investment casting techniques and vacuum induction melted alloys. There are a few solid solution nickel alloys, which find applications in heat shields and exhaust systems, but most alloys are precipitation-hardening alloys where the precipitated phase, $Ni_3(Al,Ti)$, called gamma prime (γ'), increases the high-temperature strength. Heat treatment of these alloys consists of solutionizing at 1700–2150°F to form gamma (γ), followed by one or more precipitation heat treatments at 1100–1500°F. Table 8.4 gives the nominal composition and typical mechanical properties of some common nickel-base superalloys.

8.3.2   Invar and Other Nickel-Iron Alloys

There are a number of nickel-iron alloys that display lower than normal thermal expansion, a useful characteristic for many applications such as bimetal strips, glass-to-metal seals in light bulbs or integrated circuit packages, and rods for geodetic standards. One alloy, Invar, containing 36% Ni, has little or no change in length over ordinary changes in temperature. We call these low-expansion phenomena the *Invar anomaly*, which is caused by ferromagnetic expansion of the fcc crystalline lattice. Such expansion counteracts the thermal vibrations of atoms that give rise to normal expansion behavior. The effect disappears at an inflection temperature that closely agrees with the Curie temperature. Figure 8.7 illustrates the effect of composition on the thermal coefficient of expansion at room temperature for the binary alloys.

A number of ternary alloys have been developed for special applications. For example, cobalt increases the inflection temperature of Ni-Fe alloys, but also increases the coefficient of expansion. **Kovar**, an alloy containing 29% Ni, 17% Co, and the balance Fe, is used for strong glass-to-metal seals where low-expansion Pyrex glass is required. The original electron tubes that began the electronics industry made possible by these thermally matched seals. Addition of 12% Cr to Invar produces an invariant elastic modulus in **Elinvar**. This alloy is used in tuning forks and balance wheels for clocks and watches.

There are many applications of Ni-Fe alloys that depend on their magnetic properties. For example, fifty-fifty alloys have relatively high saturation and high permeability that make these alloys useful in audio transformers, coils, and relays.

Table 8.4
Nominal composition and high-temperature properties of selected nickel-base superalloys

Alloy type	Alloy	Ni	Cr	Fe	Mo	Co	C	Nb	Ti	Al	Other	Yield strength	UTS (ksi)	Percent elongation
											Nominal composition (wt %) →	**1600°F mechanical properties**		
Cast	IN-100	60	10	—	3	15	0.18	—	4.7	5.5	1 V, 0.06 Zr, 0.014 B	101	128	6
	MAR-M 200	60	9	—		10	0.15	1	2	5	12 W, 0.05 Zr, 0.015 B	110	122	4
Wrought Solid solution	Hastelloy X	49	22	15.8	9	1.5	0.15	—	—	—	0.6 W	26	37	50
	Inconel 600	76	15.5	8	—	—	0.08	—	—		<0.25	9	15	30
	Inconel 625	61	21.5	2.5	9	—	0.05	3.6	0.2	0.2		40	41	125
Precipitation hardening	Astroloy	56.5	15	<0.3	5.2	15	0.06	—	3.5	4.4	0.03 B	100	112	25
	Inconel 718	52.5	19	18.5	3	—	0.15	5.1	0.9	0.5	<0.15 Cu	48	49	88
	Nimonic 115	55	15	1	4	15	0.2	—	4	5	0.04 Zr	80	120	18
	Rene' 41	55	19	<0.3	10	11	0.09	—	3.1	1.5	0.01 B	80	90	19
	Udimet 500	48	19	<4	4	—	—	—	3	3	0.005 B	72	93	20
	M252	56.5	19	<0.8	10	10	0.15	—	2.6	1	0.005 B	70	74	18

Figure 8.7
Linear thermal expansion coefficient at room temperature for Ni-Fe alloys

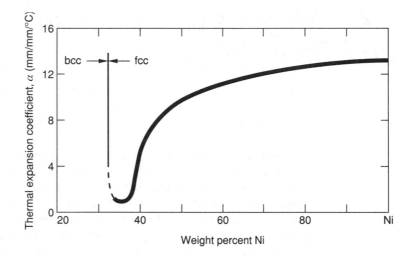

High-nickel Ni-Fe alloys have lower saturation but very high permeabilities, which has led to numerous applications in shielding undesirable electronic signals and for magnetic amplifier coils.

8.3.3 Nickel-Base Electrical Alloys

Nickel alloys containing chromium and iron have high electrical resistance and are used as electrical heating elements in such diverse applications as toasters in our kitchens to large industrial heat-treating furnaces. Electrical resistivities are high, 108 micro-ohm centimeters ($\mu\Omega$ cm) for **Nichrome**, an alloy containing 78.5% Ni, 20% Cr, and 1.5% Si, to 118 $\mu\Omega$ cm for an alloy containing 35% Ni, 20% Cr, 43.5% Fe, and 1.5% Si.

There are also thermostat and thermocouple applications for special nickel alloys. In a thermostat, two metals or a metal and nonmetal are bonded together. Differential thermal expansion causes the composite to bend as temperature is changed. Temperature changes can then be converted into mechanical energy for control, indicating, or monitoring. Although many alloys are used, some with resistance heating and some with low resistivity, Ni-Fe and Ni-Cr-Fe are among those that are commercially important.

Thermocouples indicate temperature for process control in many manufacturing operations. There are two common types, called *J* and *K*, that utilize two elements of different composition. Type J thermocouples can use **constantan**, a 45% Ni-55% Cu alloy, as the negative element and type K thermocouples frequently use a 90% Ni-9% Cr positive element and a 94% Ni alloy containing Si, Mn, Al, Fe, and Co for the negative element.

8.4   Magnesium, Titanium, and Their Alloys

Magnesium and titanium are similar in more than one way. Both are low-density elements, which is beneficial for specific strength, but both also have hcp crystal structure that makes them difficult to deform. Magnesium has a specific gravity of 1.74, but it is expensive and it is the most flammable metallic element, which makes it difficult to melt and cast because it burns readily in air. **Magnesium alloys** also have relatively low strength and poor resistance to wear and fatigue. We do consider them, however, because they are used to advantage in aeronautics and aerospace applications that place such importance on the low density (two thirds that of aluminum). Cargo shelves and racks and even some extrusions and forgings for these industries are made from magnesium alloyed mainly with aluminum and zinc.

Titanium has attributes that make it much more useful than magnesium alloys, despite its crystal structure. Titanium is also a light metal, with specific gravity of 4.54, about 56% that of steel. Its strength rivals that of steel, however, so it finds many uses in the aerospace industry because of the high strength-to-weight ratio. Its superior corrosion resistance also leads to many structural applications where corrosion can be encountered. Titanium is expensive, however, because of the difficulty in extracting the pure metal from ore, the necessity to prepare alloys by vacuum arc melting and remelting because of potential embrittlement with oxygen, and the difficulty in deforming the hexagonal crystal structure except at high temperatures.

Titanium alloys fall into three categories that are based on their microstructure. Pure titanium is hexagonal but transforms to bcc at 883°C; by alloying with aluminum, vanadium, or tin, the bcc structure can be stabilized. The three types of titanium alloys are alpha, which is hcp, beta, which is bcc, or alpha + beta. α alloys are preferred for high-temperature applications because of superior creep resistance and for cryogenic applications because of superior toughness at these low temperatures. Alloys are normally annealed and have good weldability, but are not easily forged.

Alpha + beta alloys are readily forged or extruded, with some property control possible by regulating the amounts of each phase through the processing temperature. Because the two phases are present, microstructure and properties can be controlled by cycling through the phase transformation, much like the control of microstructure of steel by cycling through the eutectoid reaction. Beta alloys are the most deformable of the titanium alloys and can be cold-rolled into sheet more readily than any other titanium alloys.

Table 8.5 lists the yield strength and ultimate tensile strength of titanium alloys of each type. The most important commercial alloy is Ti-6% Al-4% V. This alloy plus unalloyed grades account for 75% of all titanium production.

Table 8.5
Strength of titanium alloys

Alloy*	Yield strength (ksi)	UTS (ksi)	Percent oxygen (max)
Unalloyed, Gr. 1	25	35	0.18
Unalloyed, Gr. 3	55	65	0.25
Unalloyed, Gr. 4	70	80	0.40
Alpha			
Ti-5% Al-2.5% Sn	110	115	0.20
Beta			
Ti-13% V-11% Cr-3% Al	160	170	0.17
Alpha + Beta			
Ti-6% Al-4% V	120	130	0.20
Ti-6% Al-6% V-2% Sn	140	150	0.20

*Alloy designation gives nominal composition.

8.5 Zinc Alloys

Zinc is best known as an alloy element in brass. As a pure metal, it has little strength, a low melting point of 787°F, and is dense, with a specific gravity of 7.14 (remember, steel's is 7.87). Although zinc is hcp, cold deformation is not difficult because of its low melting point. However, most **zinc alloys** are cast, with the most important process being die casting. The largest consumption of zinc is for galvanizing steel by dip coating or electroplating in order to provide corrosion protection. We will learn more about this use in Chapter 11.

Die casting is a permanent mold process whereby molten metal is forced into a mold by pressure and allowed to solidify. Zinc-base die-casting alloys are popular because they offer higher strength than that of other die-casting alloys, except copper-base alloys, which have much higher melting temperatures. The zinc-base alloys can be cast to close dimensional tolerances and in thin sections; castings are easily finished by machining. Composition and properties of three commercial die-casting alloys are described in Table 8.6.

Table 8.6
Characteristics of zinc die-casting alloys

Alloy	Composition (wt %)			Yield strength (ksi)	UTS (ksi)	Percent elongation
	Al	Cu	Mg			
ASTM AG40A	4.0	0.2	0.04	35	41	10
ASTM AC41A	4.0	1.0	0.04	39	48	7
Zamak 2	4.0	3.0	0.03	—	52	8
Rolled zinc	—	—	—	—	21	50

Case Study 8.3

The Defective Bracket

The most common injuries among office workers result from falls. It is important that handrails be provided on stairways because they can prevent falls and injuries after a slip or trip occurs. Most building codes require handrails to withstand a force of 200 lb applied in any direction. When AJZ Computer recently moved into an old mill building, they carpeted the stairs and replaced all the handrails for the old stairways leading to the new office area. The installers followed the codes and fixed the wooden rails to brackets placed every 8 ft. These brackets, screwed into wall studs, were die castings made of ASTM AG40A zinc alloy plated with brass for cosmetic appearance.

Handrails provide a number of functions. They act as a guide along the stairway, provide stability for the infirm, act as a grab bar, and they should prevent falls. The handrails at AJZ appeared to be satisfactory for several years, but last week Mary was carrying some computer printouts downstairs when her high heel caught the carpet and she began to fall. When she grabbed the handrail, the bracket broke and she did fall. Now she is in the hospital with a broken leg.

Although Mary is covered under workmen's compensation, her boss was very upset and asked a metallurgist friend to look at the bracket. The failure analysis prepared by the metallurgist showed very large voids in the fracture surface, as shown in Figure 8.8. Examination of the surfaces of the voids at higher magnifications, however, showed cold shuts (see Figure 8.9) that are characteristic of pouring a metal at too cold a temperature. These, of course, caused failure because of stress concentration.

A new handrail has been installed, but you can be sure that the installers checked each bracket more carefully and added more support brackets for the rail!

8.6   *Lead, Tin, and Their Alloys*

Lead is currently a metal to be avoided because of its toxicity. We hear concerns about lead removal that are beginning to rival those voiced about asbestos removal efforts. We already have eliminated lead in gasoline and collapsible lead tubes have been largely replaced with plastic. The latter change was made for economic reasons, not because of public furor.

We do still use a lot of lead today, however, because it is dense and absorbs X rays as well as sound and because it is a necessary component in automobile batteries. Tin is used primarily as a plate for steel cans to prevent rusting, but lead containing a small amount of tin is also used for plating steel used for automobile gas tanks to provide corrosion resistance. The lead-plated steel is called terneplate.

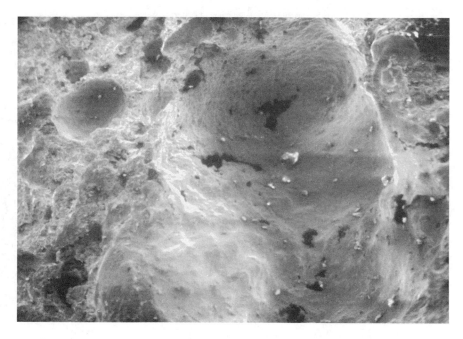

Figure 8.8
SEM micrograph of the fracture surface at the base of the bracket (26.4×)

Alloys almost always use lead and tin together. Antifriction bearing materials, or babbitts, can be either tin-base or lead-base, with the original **babbitt metal** alloy being 84% Sn, 8% Cu, and 8% Sb. Lead babbitt is more widely used for cost reasons and has provided good resistance to moderate loads at small sliding speed. It is made of the composition 85% Pb, 5% Sn, 10% Sb, and 0.5% Cu.

Perhaps the largest use of lead and tin is for solder alloys (these will be discussed more thoroughly in Chapter 10), where they find applications principally in the plumbing and electronics industries. The main solder composition is 60% Sn-40% Pb, close to the eutectic temperature in the Pb-Sn phase diagram shown in Figure 4.6.

Summary

Nonferrous metals and alloys include all metals and alloys that do not contain iron as the main constituent. In this chapter, we have examined many of the alloys whose applications arise from properties other than mechanical strength. Copper, for

Figure 8.9
Cold shuts on the surface of a large void (650×)

example, is used for its excellent electrical conductivity and ease of manufacture in many forms. Copper-base alloys such as brass, bronze, and precipitation hardenable beryllium-copper have also received broad commercial acceptance. Copper-nickel and nickel-copper alloys are used for strength and corrosion resistance. Nickel alloys are used for high-temperature strength as superalloys and as magnetic alloys for shielding of electrical fields or for electronic components. Invar, an alloy of 36% Ni with Fe, is fcc and has zero coefficient of expansion over a reasonable temperature range. Aluminum and its alloys find application because of their light density, corrosion resistance, and formability. Many aluminum alloys are age hardenable, which helps improve the poor strength of the pure metal. Other metals, such as magnesium and titanium, have few applications because of difficulty in forming them and their reactivity, which increases their cost. Lead, tin, and zinc have applications that take advantage of their low melting ranges and easy fabrication. Zinc alloys are usually die-cast. A major use of lead and tin is for solder for electronics and metal joining.

Terms to Remember

babbitt metal	magnesium alloys
beryllium-copper	Nichrome
cartridge brass	nonferrous
constantan	superalloys
cupronickel	thermocouples
die casting	thermostats
electrical contacts	titanium alloys
Elinvar	yellow brass
Kovar	zinc alloys

Problems

1. List two applications for each of the following:
    a. aluminum alloys
    b. copper alloys
    c. nickel alloys
    d. hexagonal metal alloys
    e. lead, tin, and their alloys
2. Which nonferrous metals are used in the pure, unalloyed form?
3. Which copper alloys are heat treatable? non–heat treatable?
4. Which aluminum alloys are heat treatable? non–heat treatable?
5. Compare the precipitation in superalloys to that in aluminum age hardenable alloys.
6. Why would Invar be used in bimetal strips?
7. Dumet wire has the composition 42% Ni-58% Fe. It is coated with copper and used in making incandescent light bulbs. Explain its purpose.
8. How might you control the microstructure of an $\alpha + \beta$ titanium alloy by heat treatment?
9. Why are most stepladders made of aluminum alloys or wood instead of magnesium alloys, which are lighter?
10. How would you eliminate the defects found in the zinc alloy die casting in Case Study 8.3?
11. Describe how lead and tin babbitts work as bearing materials.

9

Deformation Processing of Metals

Engineering alloys are subjected to many different processing operations, usually done to alter the properties of the metal or to change its shape. In this chapter, we will examine the role of processing in the structure-property-performance relationships that are needed to achieve the desired shapes and properties both economically and according to specifications. We will begin by looking at what happens to the metal when a force is applied to it, learn what dislocations are and how they affect metal deformation, find out what strain hardening is, discover what the differences in hot and cold working are and why recrystallization occurs, then look at some real production shaping processes.

9.1 Plastic Deformation

In Chapter 1, we learned about engineering stress and strain, where the original area, A_o, is used to calculate the tensile properties from the load applied in testing.

In order to understand what is happening in the metal when we apply an external load to it, however, it is more appropriate to utilize true stress and strain. **True stress** and **true strain** are defined as

$$\sigma = \frac{P}{A}$$
$$\text{and } \varepsilon = \frac{\Delta \ell}{\ell}$$

Engineering stress and strain can be compared with true stress and strain in Figure 9.1. For the elastic region, we find no difference, but for the plastic deformation that precedes fracture, the strength of the metal increases with strain but at a decreased rate. This means that the metal gets stronger when it is deformed.

We call the tensile forces that deform metal *normal forces* because they are applied perpendicularly to the metal. Metals can also be deformed by *shear forces* that do not lie in the same plane and therefore oppose each other. The deformation, however, is very different from tensile deformation, as shown in Figure 9.2. In this figure, we define **shear stress**, τ, as the load, P, divided by the area, A, and the **shear**

Figure 9.1
Comparison of engineering stress and strain and true stress and strain

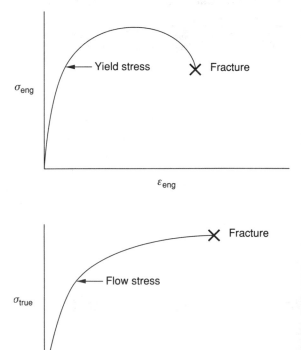

Figure 9.2
Deformation caused by shear force *P*

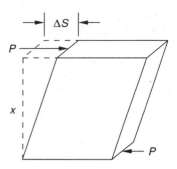

strain, γ, as the distortion, ΔS, divided by the distance, x. Shear deformation can be elastic or plastic and we can describe elastic behavior by the following relation:

$$\tau = G\gamma$$

where G is the shear modulus. We call this relation Hooke's law for shear.

We need to understand both tensile and shear forces because the real forces that deform metals can be described or resolved into the normal and shear components.

9.1.1 Slip

Plastic deformation is permanent deformation that occurs at stresses greater than the yield strength. We know from microstructural studies that metals deform by **slip** of atomic planes (called *glide planes*) within each grain, such slip occurring preferentially in the closest packed crystal planes and in the closest packed crystal directions. Slip is observable microscopically in surfaces that have been polished and then deformed, as shown in Figure 9.3. Note in this figure that slip lines never cross grain boundaries. Although slip lines are straight in fcc metals, they are wavy in bcc and hcp metals. It is easy to understand why slip occurs on the densest packed planes and densest packed directions because these planes and directions represent the smoothest ones for slipping or sliding two planes across each other, much like dealing a new deck of cards. The new cards are much smoother than old, much used ones, therefore, they slide across each other more easily than do well-worn cards.

The combination of a slip plane and slip direction is known as a **slip system**. We know that plastic deformation can occur without breaking and that stress is continuous, that is, it affects all grains, not just individual grains. This was first explained by von Mises, who showed that changing shape without changing the volume can only occur if slip occurs in at least five slip systems. Most metals that we normally produce are fcc or bcc, which have sufficient slip systems to be deformed. However, hcp metals such as titanium, magnesium, and zinc have only three slip systems active at room temperature. These hcp metals can be economically deformed only at higher temperatures where secondary slip systems are activated and a second type of deformation called *twinning* occurs. We need only be concerned with understanding slip, however, because of its primary importance in deformation.

Figure 9.3
Slip lines in Cu-2% Al alloy
(850×)
(From W. Hayden, W. G. Moffatt,
and J. Wulff, *The Structure
and Properties of Materials,*
Vol. 3: *Mechanical Behavior,*
John Wiley & Sons, 1965.)

Our understanding of the forces that are involved in slip comes from studies of single crystal deformation and the ability to resolve the forces into shear and tensile forces, as demonstrated in Figure 9.4. It is the shear force that is resolved onto the slip plane that causes slip to occur. When slip first happens, the stress to cause it is termed the *critical* **resolved shear stress**, τ_c.

The value of the critical resolved shear stress is given by Schmid's law:

$$\tau_c = \sigma(\cos \lambda)(\cos \phi)$$

τ_c has been measured for single crystals of different metals and different orientations and is approximately equal to $10^{-4}G$, where G is the shear modulus. The experimental values for critical resolved shear stress are significantly below values determined theoretically by considering atomic bonding forces.

9.1.2   Dislocations and Slip

We now know that the discrepancy between theory and measurement comes about because of the linear dislocation defects that were described in Chapter 3. Our understanding of deformation requires knowing that dislocations represent energy, the manner in which dislocations move under an applied stress, and how the number of dislocations can change.

If we consider an edge dislocation, the extra plane of atoms causes elastic stress and strain in the atomic bonds surrounding the dislocation along its entire

Figure 9.4
Schmid's law for critical resolved shear
stress

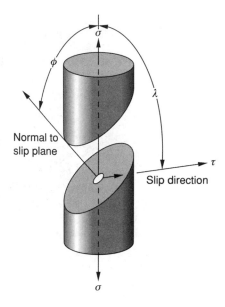

Normal to
slip plane

Slip direction

length, as depicted in Figure 9.5. Elastic energy is associated with these stresses and strains. We need to know only that this energy exists and is a function of the number and length of all dislocations present. Remember that equilibrium is represented by the lowest energy state and that higher energies provide a driving force toward equilibrium. We'll come back to these energies and show how important they are later in this chapter, but now we must look at how dislocations move under stress and how that movement makes slip occur more easily.

The simplest possible motion is that of an edge dislocation, such as that illustrated in Figure 9.6a, moving to the right in a single grain under the influence of the shear stress, τ. The stress moves atom A closer to atom C than to atom B and moves atom C closer to atom A than to atom D. The work in separating atom C from atom D is reduced by the energy released because atom A and atom C are nearer to their equilibrium positions. The net effect is movement of the dislocation to the right with reduced stress required. As the stress is continued, the dislocation continues to move to the right until it moves out of the grain, creating one unit of slip (Figure 9.6d).

Dislocations can be resolved with electron microscopy using either etch-pit techniques or transmission through thinned films; the latter is demonstrated in Figure 9.7. From studies such as these, we have learned that dislocation densities for well-annealed polycrystalline metals is 10^7–10^8 lines/cm^2, but is 10^{11}–10^{12} lines/cm^2 for heavily cold worked metals. Some mechanism, therefore, must exist that can produce dislocations during deformation. In manufacturing, we need not worry about the actual mechanism, but we do need to consider the consequences. As the number of dislocations increases, their interactions also increase, impeding or slowing dislocation motion. The stress necessary to move dislocations increases, which means that the metal becomes stronger. We call this phenomenon **strain hardening**.

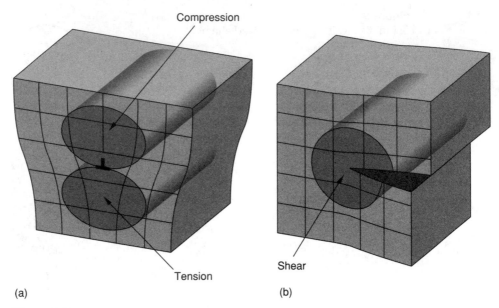

Figure 9.5
Stress and strain fields surrounding (a) an edge and (b) a screw dislocation

Metals characteristically strain-harden at a diminishing rate, that is, the increase in stress decreases as the strain increases. Because of this, the plastic region of the true stress–true strain curve (Figure 9.1) can often be approximated by

$$\sigma = K\varepsilon^n$$

where K is a strength coefficient and n is the strain-hardening exponent. For a perfectly plastic material, n is zero and can increase to one for perfectly elastic behavior. If we want to deep-draw a cup-shaped metal, we would want the metal to flow with little strain hardening (a low value of n), but if we want a structural metal with high ultimate–to–yield strength ratio, then we would select a metal with a high value of n. We will see that the higher the value of K and the higher the strain-hardening coefficient, the more difficult processing the metal becomes. For most metals, n has a value between 0.10 and 0.50. A sampling of some strain-hardening values for common metals appears in Table 9.1.

9.1.3   Stored Energy of Cold Work, Recovery, Recrystallization, and Grain Growth

If we recall that the dislocation density increases with cold work and that energy is associated with each dislocation, then we recognize that there must be increased

Figure 9.6
Slip created by movement of edge dislocation caused by applied shear stress

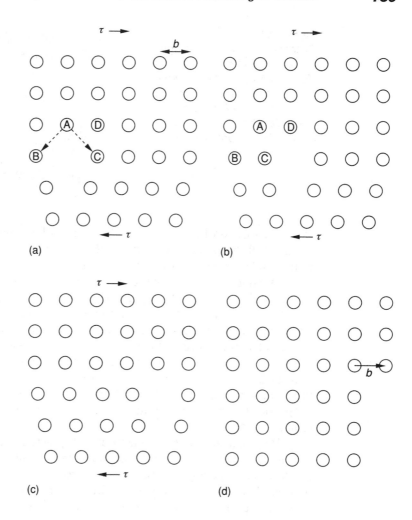

energy in cold-worked metals. This increased energy is called the *stored energy* of cold work. Although only a small percentage of energy expended in deforming a metal is actually stored (see Figure 9.8), its importance to processing cannot be overemphasized. Because of this stored energy, we can control processing and the properties of the final shape.

When we heat a metal that has been strain-hardened, energy is released. This can happen as a function of time in an isothermal heat-treating process or as a function of temperature in continuous heating, that is, an anisothermal heat-treating process. The energy is released in two stages, the first at lower temperature (or shorter time), which only affects the most structure-sensitive properties, and the second at higher temperature (or longer time), which alters both properties and microstructure. Figure 9.9 shows the changes occurring in energy, electrical resistiv-

Figure 9.7
Dislocations in niobium single crystal (11,600×)
(From W. Hayden, W. G. Moffatt, and J. Wulff, *The Structure and Properties of Materials,* Vol. 3: *Mechanical Behavior,* John Wiley & Sons, 1965.)

ity, and hardness of nickel that has been strain-hardened and heated anisothermally. Figure 9.10 shows the microstructural changes in low-carbon steel reduced 50% by rolling and then heated for 10 minutes at 970°C.

During the major energy release, we see a complete restoration of the mechanical and physical properties of the metal altered by the deformation. The release occurs simultaneously with the formation of a new set of strain-free grains, a phenomenon we call **recrystallization**. Recrystallization is preceded by the partial release of stored energy, which we refer to as *recovery*. In manufacturing, recovery is most common in stress relief anneals that are designed to remove residual stresses, such as thermal stresses associated with differential cooling or heating. Recrystallization plays a very important role in the processing of metals, either in restoring properties to enable additional processing or in controlling the properties of the final product.

Recrystallization is a nucleation and growth process brought about by the driving force of the stored energy of cold work. The temperature or time when energy is released is dependent on this driving force; therefore, the more severe the amount of strain hardening, the larger the driving force and the lower the recrystallization temperature or time. Figure 9.11 illustrates the effect of temperature and time on the recrystallization of heavily deformed copper.

Table 9.1
Strain-hardening values for selected metals

Metal	Condition	*n*	*K* (psi)
0.05% C steel	Annealed	0.26	77,000
SAE 4340 steel	Annealed	0.15	93,000
0.6% C steel	Quenched and tempered 1000°F	0.10	228,000
0.6% C steel	Quenched and tempered 1300°F	0.19	178,000
Copper	Annealed	0.54	46,400
70-30 brass	Annealed	0.49	130,000

Figure 9.8
Stored energy of cold work

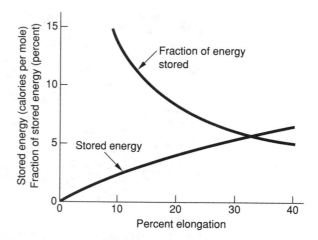

Once recrystallization has occurred, the stored energy of cold work has been reduced. However, there is still a lower energy state for equilibrium that is achieved by grain growth, where some grains grow at the expense of others by lowering the surface energy associated with grain boundaries. In manufacturing, grain growth is controlled by the time and temperature of the final **process anneal**. It is just as important to limit the amount of grain growth prior to deformation, however, because the recrystallization temperature is also dependent on grain size prior to deformation.

Figure 9.9
Change in properties of cold-worked nickel associated with energy release during anisothermal heating

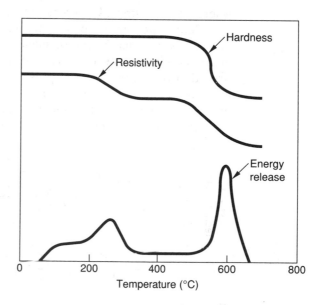

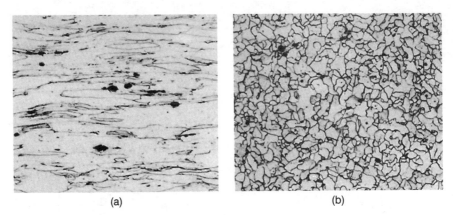

(a)   (b)

Figure 9.10
Low-carbon steel microstructure (100×): (a) cold-rolled 50%, RB 82 hardness;
(b) cold-rolled 50%, recrystallized, RB 59 hardness
(From L. E. Samuels, *Optical Microscopy of Carbon Steels,* ASM International, 1980.)

Another reason that we must control the grain size of annealed metals is that the strength is affected. At low temperatures, grain boundaries are stronger than the grains themselves, which is thought to be a result of barrier action to slip within the grains. Therefore, at these low temperatures (which include room temperature), there is a relationship between yield stress and grain size that is known as the **Hall-Petch relation**:

$$\sigma = \sigma_o + kd^{-1/2}$$

where σ is yield stress, k is a constant, and d is the grain diameter.

Many applications require strength that is not achievable with annealed material. This is particularly true of single-phase (or nearly single-phase) alloys such as austenitic stainless steel, low-carbon steel, and α brass. In these alloys, strength is provided by controlling the plastic deformation processing, termed the *cold work–anneal cycle*. The strength is referred to as **temper** and is obtained by controlling the amount of deformation after the last process anneal. Table 9.2 provides an example of the properties of 70/30 α brass for different tempers.

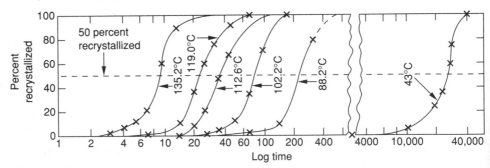

Figure 9.11
Isothermal recrystallization curves for pure copper (99.999% Cu) cold-rolled 98%.
(Reprinted from B. Decker and D. Harker, Trans. AIME, *188,* p. 887, 1950.)

Table 9.2
Temper designation and properties of 70/30 alpha sheet brass

Description	Percent reduction in area after anneal	Yield strength (ksi)	UTS (ksi)	Percent elongation
Annealed				
¼ hard	9.9	40	54	43
½ hard	20.7	52	62	23
Hard	37.1	63	76	8
Extra hard	50.1	65	86	5
Spring	60.5	—	94	3
Extra spring	68.6	—	99	3

9.1.4 Texture in Deformation Processing

Texture, or preferred grain orientation, can be developed during deformation processing of polycrystalline metals. Such texture is easy to understand knowing that crystals deform on specific slip systems. As the amount of cold reduction increases, so does the crystallographic alignment of grains. Of course, such texture produces directional properties (anisotropy) that can be either beneficial or detrimental, depending on the application. Because this is true, we have to know or specify whether texture can be tolerated in the application because strict process control must be exercised to provide the required properties in the final stage of processing. For example, a rule of thumb limits cold work to about 65%, followed by a process anneal. This is particularly critical in the cold work–anneal cycles because once deformation texture occurs, it cannot be removed. (Texture is altered by recrystallization but not eliminated.)

Preferred orientation is developed by compressive forces during cold deformation, but small amounts of alloy elements and the prior grain size before cold working (which we call **penultimate grain size**) can either increase or decrease this development. The most common example of the importance of texture is in the **deep drawing** of sheet metal whereby a flat sheet is formed into a finished shape by a combination of drawing and stretching, using a punch and die. For good stretching characteristics, a high strain-hardening exponent, n, is desirable. We can control the value of n mainly through grain-size control. Larger grain size is preferred, but there are practical limitations because surface imperfections can be encountered. The maximum grain size allowable to avoid *orange peel*, aptly named to describe such surface imperfections, is about ASTM No. 6. In idealized drawing, no deformation takes place over the nose of the punch; therefore, the capability of the metal to withstand drawing depends on its ability to deform in the plane of the sheet and its ability to resist deforming in the thickness of the sheet. We call the ratio of these qualities, the plastic **strain ratio**, r, which is influenced by the amount of cold work, as shown in Figure 9.12. Other texture-related factors also affect deep drawing, such as nonuniform extension or *earing*, which occurs with cold work greater than about 80%.

Figure 9.12
Influence of cold work on
plastic strain ratio

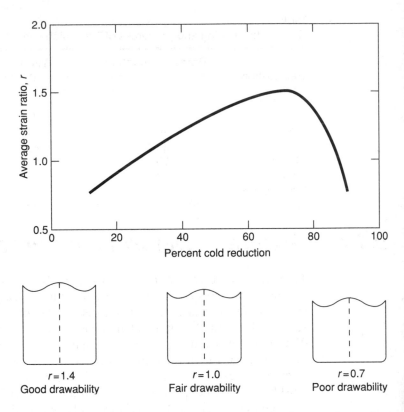

r=1.4
Good drawability

r=1.0
Fair drawability

r=0.7
Poor drawability

In electrical steel and iron-nickel magnetic alloys, texture plays an important role in achieving the properties desired. Although the iron-silicon electrical steels are bcc and the iron-nickel alloys are fcc, their magnetic properties are optimized only when the phenomenon of **secondary recrystallization** occurs. In oriented electrical steels, texture is developed to take advantage of easy magnetic directions. It is produced by 75% cold reduction, annealing at a low temperature, reducing further another 50%, decarburizing at about 700°C in dry hydrogen, then heating at 1200°C to form the secondary recrystallized structure. In the iron-nickel alloys, all that is necessary to develop secondary recrystallization is a fine penultimate grain size, cold reduction greater than 90% with final thickness of 0.015 in. or less, and a high temperature anneal, above about 900°C. In these alloys, the secondary grain size is extremely large, visible to the naked eye after etching, as shown in Figure 9.13.

Case Study 9.1

A Practical Problem of Anisotropy

A local company that manufactures pressure transducers with guaranteed performance recently found an inexplicably large increase in rejections based upon excessive mechani-

Figure 9.13
Secondary recrystallization
of Fe-49% Ni U-shaped
sheet specimen (4×)

Secondary grain

Primary recrystallized grains

cal hysteresis (strain lags behind stress for increasing or decreasing applied stress). These parts were machined from 0.065-in. thick 17–4 PH stainless steel, heat-treated and brazed into the assembly. The only purchase specifications were for certified chemical analysis. Conventional metallurgical tests, including chemical, microhardness, and microstructural analyses were not successful in finding the cause. A difference in microstructure of acceptable and rejectable parts was observed in a scanning electron microscope but could not be correlated with the presence or the lack of hysteresis. Because hysteresis can also be influenced by texture, X-ray pole figures were run for the acceptable and rejectable conditions. These pole figures clearly showed that, compared to acceptable material, the rejected material was heavily textured.

The company stopped using the heavily textured sheet material and has had no further problems. They are adding a processing specification to the chemical certification for future purchases.

9.2 *Fundamentals of Metalworking*

Useful metal shapes can be generated in three basic ways: casting liquid metal to shape, metal removal, and plastic deformation processes. Although we have been examining deformation processing, we're going to look at it from a slightly different perspective now. **Metalworking** can be simply defined as a process whereby metal is displaced from one location to another while the volume and mass of the metal are conserved, that is, not changed. From the previous discussion, we also recognize that the metalworking process must be controlled in order to ensure desired properties. It is generally true that the properties of wrought or plastically deformed metals are superior to those of cast metals because porosity and segregation are eliminated and cast microstructure is refined during metalworking.

9.2.1 *Deformation Processes*

Many processes exist for specific metalworking practices, but we categorize them according to the application of forces to the workpiece. There are five specific metalworking processes:

1. direct compression processes
2. indirect compression processes
3. bending processes
4. shearing processes
5. tension processes

These categories are self-explanatory, with the exception of indirect compression processes. In these, the primary forces that are applied are tensile forces, but the deforming forces are compressive. Examples of all of these metalworking processes appear in Figure 9.14.

We sometimes find it convenient also to classify metalworking practices by their function. Accordingly, a **primary process** can be defined as one that will reduce an initial shape to one that is suitable for further processing, and a **secondary process** is one that begins with an intermediate shape and converts it to the final desired shape. This classification is not precise, however. Although a bending or shearing operation will almost always be a secondary process, direct or indirect compression processes often can fit either description. Tensile processes are rarely used because the metal cannot be easily constrained to permit large deformations.

In order to understand what occurs during any metalworking process, it is helpful to view the process as a total system, such as that depicted in Figure 9.15. The shape is actually changed in the deformation zone, but we also have to consider

Figure 9.14
Examples of metalworking categories: (a) direct compression, (b) indirect compression, (c) bending, (d) shearing, (e) tension

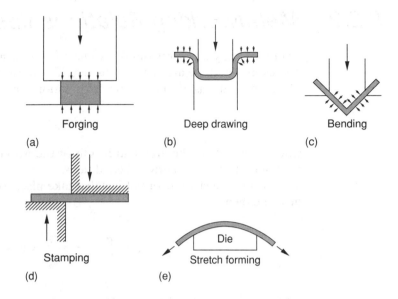

the distribution of stresses and strains, and metal velocities. Applied forces must be sufficient to deform the metal plastically but not fracture the workpiece. Strain hardening, recrystallization, and fracture are all important, but for the specialized conditions of high strain rates and/or elevated temperatures. Flow stress is strongly influenced by the strain, strain rate, and temperature in the constrained deformation zone and it is often the case that we cannot reproduce these conditions in laboratory scale studies.

In all metalworking systems, the workpiece is in intimate contact with nondeforming (elastic) tools, and friction along the interface between workpiece and tool necessitates lubrication. If not taken care of, friction can cause tool wear and poor surface finish of the product.

Figure 9.15
Deformation processing system

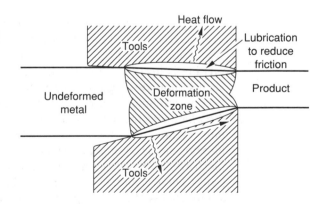

9.2.2   Metalworking Relationships

There are only a few relationships that we use in manufacturing by metalworking processes, but these are used extensively. The most fundamental is the conservation of volume and mass during plastic deformation. This is described by

$$A_f \ell_f = A_o \ell_o$$

where A and ℓ are the area and length of the workpiece and the subscripts f and o represent the final and original conditions.

There are large deformations that take place, so we use the total true strain to describe them:

$$\varepsilon = \int_{\ell_o}^{\ell_f} \frac{d\ell}{\ell} = \ln \frac{\ell_f}{\ell_o} = \ln \frac{A_o}{A_f}$$

Stress analysis of metalworking processes is very complex because it must consider the effects of friction, velocity, and plastic deformation constraints as well as stress and strain. However, the following empirical relationship is useful in controlling processes such as extrusion or wire drawing:

$$\sigma = \frac{P}{A} = K \ln \frac{A_o}{A_f}$$

where K is known as the extrusion constant. However, K is affected by temperature, die changes, lubrication changes, and so on.

9.2.3   Hot Working

The beginning metal shape for the first deformation process is always a cast metal. As we learned in Chapter 5, cast microstructures inherently contain porosity, segregation, and coarse columnar grains that limit the properties of these materials. **Hot working** is invariably the initial step in deforming mass-produced metals, mainly because hot working provides reduced resistance to deformation; this lowers the force requirements on equipment and permits larger deformations. Hot working also eliminates the deleterious characteristics of cast metals. Rapid diffusion at high temperatures plus the atomic displacement of deformation slip reduce chemical inhomogeneities, pores are welded together by the heat and pressure, and the cast microstructure is broken down and replaced with more equiaxed recrystallized grains. These changes due to hot working all tend to increase the strength, ductility, and toughness of wrought products compared to cast products.

There are certain disadvantages to hot working. Because high temperatures are usually involved, the metal surfaces oxidize, resulting in material loss or entrapment as *rolled-in oxide*. Surface decarburization of steels can also be a problem, frequently requiring extensive surface refinishing to remove the decarburized layer. In addition, because we must allow for expansion and contraction, the dimensional tolerances for hot-worked mill products are greater than for cold-worked products. Furthermore, the structure and properties of hot-worked metals are generally not so uniform as in metals that have been cold-worked and annealed because the deformation is always greater in the surface layers than in the interior, leading to finer recrystallized grains in the surface. Also, interior grains will cool more slowly, leading to increased grain growth for the interior.

The lower temperature limit for hot working is the lowest temperature at which the rate of recrystallization is fast enough to eliminate strain hardening in the time that the metal is at temperature. For a given metal or alloy, the lower hot-working temperature will depend on such factors as the amount of deformation and the time the metal is at temperature. Since the greater the amount of deformation the lower the recrystallization temperature, the lower temperature limit for hot working is decreased for large deformations. Metal that is deformed rapidly and cooled rapidly will require a higher hot-working temperature for the same degree of deformation than will a metal slowly deformed and slowly cooled.

The upper limit for hot working is determined by the temperature at which either melting or excessive oxidation occurs. Generally, we limit the maximum hot-working temperature to 90°F below the solidus or melting temperature. Even a thin film of a lower melting point constituent at a grain boundary will make the metal crumble into pieces when it is deformed, a condition referred to as *hot shortness*. (See Chapter 7. This is the reason we add manganese to all steels, because it forms an innocuous inclusion with sulfur and sulfur lowers the melting point of iron dramatically.)

Most hot-working operations are carried out in a number of multiple passes or steps. Generally, the intermediate passes are done at temperatures well above the minimum working temperature in order to take advantage of the economics offered by lower flow stress. It is likely that some grain growth will occur subsequent to recrystallization at these temperatures. For the last pass, therefore, the working temperature is lowered to the point at which grain growth will be negligible to ensure the fine penultimate grain size usually preferred. The finishing temperature is usually just above the minimum recrystallization temperature. Also, in order to ensure this fine penultimate grain size, we always make the last pass relatively large.

9.2.4 Cold Working

Cold working of metals includes all deformation processes conducted below the recrystallization temperature of the metal. Such processes are best considered as secondary processes because they rarely begin with cast microstructures. This is true because of the large deformation forces required to deform a large section and limitations of power in available machinery. Oxidation is not a serious problem and thermal expansion and contraction are much lower than for hot working, so cold-worked products have better tolerances and better surface finishes than hot-worked products.

In order to produce superior surface finishes, we must condition the hot-worked starting material prior to cold working. Proper conditioning by pickling with chemicals or grinding removes the scale that forms on hot-worked metals. Grinding can also remove surface cracks that might form laps or propagate, thus reducing yield or causing failure during cold working if not removed. If conditioning is inadequate or not done at all, we can find defects such as rolled-in oxides and internal microcracks in the final product.

As with hot working, there are disadvantages to cold working. Because we carry out the deformation below the recrystallization temperature, strain hardening occurs and texture can develop. Consideration of these effects must be made for each metal processed so that we can tailor cold work–anneal cycles for individual products. Specific examples of such planned cold work–anneal cycles will be given under the specific deformation processes that follow.

9.3   Forging

Forging is the oldest of metalworking processes, having its origins in primitive blacksmithing. It is a direct compression process whereby a metal is hammered or pressed into a useful shape. *Forged* has traditionally been the mark of quality for hand tools and hardware. Pliers, hammers, wrenches, cutlery, and even fine surgical instruments are common examples of forged items. Many other products in which uniform properties are required, such as crankshafts, axles, eye bolts, and machinery parts, are also forged.

The two broad categories of forging are open-die and closed-die forging. We carry out open-die forging between two flat dies or dies of simple shape when we have large workpieces or when the number of parts is small and we intend to finish machine them. We often use open-die forging to preshape metal for closed-die forging. The simplest open-die forging is called **upset forging**, which is commonly used to break down the cast microstructure of ingots, forming shaped **billets** for secondary processing. We homogenize, or *soak*, the ingots at high temperatures and utilize manipulators, which are powered vehicles equipped with hydraulic grips that can clamp and rotate the workpiece within the dies. Reheating is usually needed before the final billet shape is completed.

The compression that occurs in upset forging is shown schematically in Figure 9.16. The distortion, or *barreling*, that can occur is the result of friction between the workpiece and dies, chilling of the workpiece by the cooler dies (which increases deformation resistance), and inhomogeneous deformation. This inhomogeneous

Figure 9.16
Compression of a cylindrical
workpiece between flat dies:
(a) beginning ingot shape,
(b) intermediate shape,
(c) final billet shape

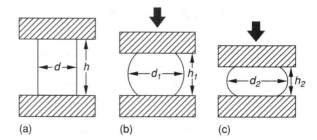

(a)　　　　　(b)　　　　　(c)

deformation is the result of differences in the actual stress applied across the work-piece. Stress analysis has demonstrated a peak in the applied stress at the center of the workpiece that, in effect, acts like a cone compressing the workpiece. The end result is macrostructure such as that shown graphically in Figure 9.17. Area I in this figure is shaped by the vertical compression caused by the force cone combined with little lateral motion because of friction. Area III is the result of center material moving radially outward and the bulk of the deformation is concentrated in the center, area II.

Stress analysis also has shown us that the forging force increases exponentially as the workpiece height decreases. This is particularly important to us when it comes to closed-die forging where we might want to reduce weight by introducing thin webbed sections. In **closed-die forging**, we use carefully matched dies to make products with close dimensional tolerances. These dies are expensive and large production runs are necessary to justify their costs. Three die sets are normally employed. A preshaped billet is first forged in a *blocking die*, where the greatest change in shape takes place. The second die is a *finishing die*. We use slightly more metal than is needed to fill the finishing die, causing the metal to exude between the dies as flashing. This flashing regulates the escape of metal and increases the flow resistance of the die system, assuring that metal fills all recesses because of internal pressure buildup. In the last die set, a *trimming die*, the flashing is removed.

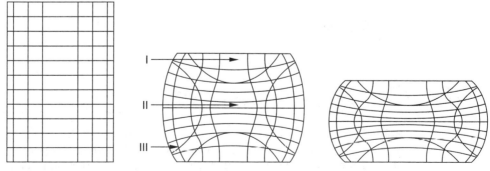

Figure 9.17
Inhomogeneous deformation in a forged cylinder

9.3.1 Forgeability

Forging has always been a blend of art and technology, requiring many tests, designs, and modifications before production can begin. Decision making involves considering both product and tool design, die materials, and lubricants in addition to the processing itself. From the materials viewpoint only, we find the forgeability to be related to the high-temperature strength and to oxidation resistance. There are a number of tests to determine forgeability, but the most widely used is the upset test. This test consists of upset-forging a series of cylindrical billets to various thicknesses, or to the same thickness, but with varying length-to-diameter ratios. The limit for upset forging without failure by radial or peripheral cracking is considered the measure for forgeability. The following common engineering metal alloys are listed in order of decreasing forgeability:

> aluminum alloys
>
> magnesium alloys
>
> copper alloys
>
> carbon and low-alloy steels
>
> ferritic and martensitic stainless steels
>
> austenitic stainless steels
>
> nickel alloys
>
> titanium alloys
>
> iron-base superalloys
>
> cobalt-base superalloys
>
> niobium alloys
>
> tantalum alloys
>
> molybdenum alloys
>
> nickel-base superalloys
>
> tungsten alloys
>
> beryllium alloys

9.4 Rolling Processes

Rolling is a direct compression process in which plastic deformation takes place by passing a workpiece between cylindrical rolls. This is the most popular metalworking process because of the capability for large production to close tolerances. The metal is subjected to high compressive stresses from the squeezing action of the rolls and to surface shear stresses as a result of friction between the workpiece and the rolls. These frictional forces are necessary, however, because they are responsible for drawing the metal into the rolls. Rolling can be either a primary process, as in hot rolling to bloom an ingot or to produce coiled *hot band* for further cold rolling, or a secondary process, as in producing such items as household aluminum foil.

A rolling mill consists of the rolls, bearings, housing, and power drive to control the speed and pressure of the rolls. Most rolling mills fall into the category of two-high rolls for initial hot rolling, with roll widths of 24 in. to 96 in. In cold rolling, power can be reduced and surface finish can be improved at the same time by using smaller diameter rolls. Four-high rolls and cluster rolls are used to roll thinner sections such as foil. These mills require large backup rolls to prevent deformation of the rolls themselves. (In two-high mills, rolls are crowned to minimize roll distortion and ensure flat sheet across its width.) Figure 9.18 illustrates the types of rolling mills.

The main parameters that we must be concerned with in rolling are

- roll diameter, which controls the contact area where deformation takes place
- friction between the rolls and the workpiece that changes as the thickness is reduced by the deformation
- deformation resistance of the metal being rolled, including strain hardening
- presence of front or back tension in the sheet. This is used to reduce the roll pressure required and enable us to roll thinner sheet on a specific mill. It is common practice to use front and back tension and coiling both with hot and cold rolling.

In rolling, the forces reduce the thickness and extend the material only in the forward direction; therefore, the reduction in thickness is the same as the reduction in area. Successful rolling practice requires careful balance of temperature, amount of reduction per pass (bite), process annealing, conditioning of hot band, lubrication, and roll surface condition. Scheduling must consider the final sheet properties, permissible tolerances, and degree of texture allowable. Process anneals should be performed whenever possible when there is a change from one mill to another to

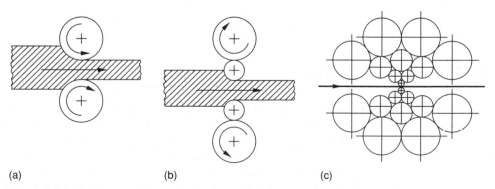

(a)   (b)   (c)

Figure 9.18
Rolling mill configurations: (a) two-high mill, (b) four-high mill, (c) cluster mill

maximize productivity. These process anneals are normally conducted by continuous passing of uncoiled sheet through a furnace. Recrystallization is ensured by controlling the speed of the sheet and the temperature and length of the furnace. In high-volume production materials, rolls can be lined up for multiple passes with no coiling in between passes.

Sample Problem 9.1

Producing 0.015-in. Sheet Steel

As production manager, you must devise a schedule to produce 0.015-in. thick sheet steel from a billet that is 4.5 in. thick, 28 in. wide, and 6 ft long. Half of the sheet must be annealed and half must be ½ hard temper (20.7% reduction after last anneal). Any texture is to be avoided. Your hot-rolling mill can produce 0.25-in. hot band and you have a two-high mill usable to about 0.062-in. sheet, a four-high mill usable to about 0.032-in. sheet, and a cluster mill to the final size.

Solution

Hot rolling.   Reduce the billet to 0.25 in. thickness, making the last pass the heaviest to ensure a fine penultimate grain size. Length is 98 ft.

Condition.   Remove oxides by pickling in chemical bath and smooth any cracks by grinding.

Two-high mill.   Reduce to 0.088 in. (64.8% cold work), process strip anneal, and reduce to 0.062 in. Length is 435 ft.

Four-high mill.   Reduce to 0.032 in. (63.6% cold work), process strip anneal. Length is 844 ft.

Cluster mill.   Reduce to 0.019 in. (40.6% cold work). Length is 1421 ft. Divide into two lengths, each 79.5 ft. Reduce one length to 0.015 in. (53.1% total cold work). Strip anneal both pieces. Annealed piece at 0.015 in. is completed; its length is 900 ft. Reduce 0.019-in. sheet to 0.015 in. (21.0% cold work); this is ½ hard and its length is 900 ft.

9.5   Extrusion

Extrusion is a direct compression process whereby a workpiece is forced through a die under high pressure. In general, we extrude cylindrical rod or tubular shapes, but we can also produce irregular shapes, particularly in soft metals such as aluminum. For example, aluminum gutters are extruded to length at the site of installation. We can also extrude these soft metals by impact at room temperature. However, we ordinarily think of extrusion as a hot-working process that requires high pressures.

The types of extrusion include direct and indirect extrusions, hydrostatic extrusion, and pierce-and-extrude techniques to produce seamless pipe (depicted in Figure 9.19). Although we do extrude to final shapes, we generally consider extrusion as a primary process, with secondary processing by operations such as wire or tube drawing to follow. Extrusion is particularly effective in primary breakdown of ingots of hard-to-deform metals, such as stainless steel, because the high compressive stresses involved reduce any tendency for cracking.

Extrusion presses are usually hydraulic and can be either horizontal or vertical. Vertical presses, of course, require less space, but are limited to indirect extrusion. Ram speed is important and can vary from a few inches per minute (below which vibrations interfere) to about 1500 inches per minute if hydraulic accumulators are incorporated. Billets or ingots are preheated to a temperature usually referred to as

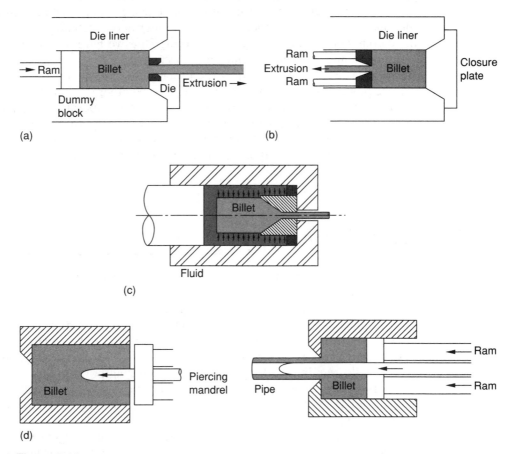

Figure 9.19
Extrusion methods: (a) direct extrusion, (b) indirect extrusion, (c) hydrostatic extrusion, (d) pierce and extrude

the extrusion temperature. The real temperature of extrusion, however, is not determinate because it depends on the heat generated by deformation and friction that varies by the preheat temperature of the die liners, the reduction, the lubrication system, ram speed, and so on.

The force for extrusion is given by the relation

$$P = A_o K \ln \frac{A_o}{A_f}$$

where K is the extrusion constant. Force is considered a function of temperature, but does depend to a lesser extent on the ram speed, lubrication, and other factors. The force in direct extrusion increases with ram travel until extrusion through the die begins. This peak value is called the *breakthrough pressure* and must be achievable to prevent stalling the billet; the force drops to a somewhat lower level then, known as the *running pressure*, until extrusion is completed. Stalling a billet is costly because the billet is somewhat smaller than the die liner and is actually upset-forged to fill the die liner before breakthrough can occur. Thus we have an appreciable removal and remachining problem if stalling does occur.

Extrusion practice can affect the properties of the product as well; for example, the properties of titanium alloys depend on whether or not extrusion takes place in the high-temperature single-phase or the lower-temperature two-phase region. Precipitation can also take place in age-hardening alloys when cooling from the extrusion temperature, but the cooling rate will vary from billet to billet because of the different extrusion conditions. Uniformity requires heat treatment.

9.6 Wire Drawing

Wire drawing is an indirect compression process whereby a reduced section is gripped and pulled through a die. The elastic die compresses the metal being drawn through it. We normally consider wire or rod drawing as a secondary process, but some forming can follow. We refer to rod drawing or wire drawing because of the flexibility exhibited by the material. Rods are too rigid to be coiled whereas wire is flexible enough to be coiled or wound onto a spool for ease of handling. Drawing processes do require auxiliary equipment, including pointing machinery (usually swaging), effective gripping means for pulling the rod or wire, and miscellaneous handling equipment such as cranes for rods and coils, pay-off stands for coils, and spooling equipment.

Central to wire and rod drawing is the die, shown in Figure 9.20, which must be able to withstand temperatures in excess of 300°C at the bearing surface and be wear resistant to ensure good surface finish and accurate dimensions. Die materials

Figure 9.20
Wire-drawing die

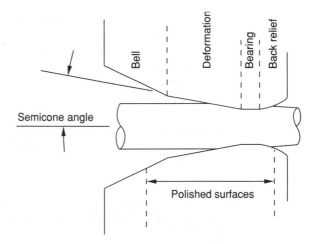

include plated alloy steels for the largest sizes where the cost is highest but the contact surface area is smallest (thus reducing wear). Carbide inserts are used for intermediate die sizes and diamond dies for smallest diameters where contact surface area is highest and wear is the biggest problem.

Drawing studies have shown that an optimum die semicone angle exists where high angles require additional forces for the high reductions and lower angles lead to higher frictional forces because of the larger contact surface area. Lubrication is absolutely necessary, with a coating initially applied to the material by dipping and lubricant dispensed onto the material immediately in front of the die during the drawing process. Any increase in friction, reduction, and bearing length increases the drawing force required.

Productivity is determined more by the ability to handle the material than by the speed of drawing. The factors that are most important are the reduction per pass, which is typically in the range of 9% to 45%, and the ability to incorporate multiple passes. For **draw benches** used in drawing straight rods, space requirements and material handling procedures are the most demanding. Once we can coil the product, a machine called a **bull block** is used. Space requirements are reduced because the product is coiled, but drawing time is increased with each reduction. Once the straight rod is drawn through the die, it is taken up and accumulated on a **capstan**, a tapered rotating drum that also provides the drawing force. At smaller diameters, only several wraps of the wire are taken up on the capstan, then it is removed and spooled separately. Tandem processes and multiple-die drawing processes, such as those shown in Figure 9.21, have been devised to improve the productivity of wire drawing.

Noncylindrical shapes are usually reduced in round dies, with final processing by forming (see Section 9.7) and the last pass by drawing through the special noncylindrical die. Seamless tubing is drawn from pipe made by the pierce-and-extrude

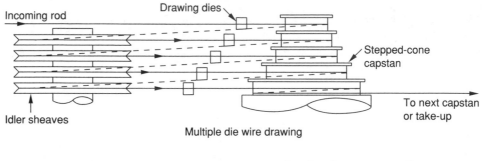

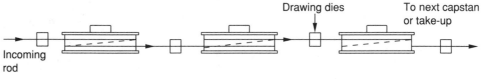

Multiple die wire drawing

Tandem drawing

Figure 9.21
Methods to improve productivity of wire drawing
(Courtesy of AT&T.)

technique using a fixed **mandrel** for straight lengths or a floating mandrel for smaller sizes that are coiled. Figure 9.22 illustrates these methods.

Wire drawing cold-works the metal and process anneals often are needed to soften it and prevent fracture. In process anneals, uniform microstructure dictates the need for conveyorized processing using pay-off and take-up reels. If a batch anneal is attempted for spooled wire, the outer wraps experience grain growth before sufficient heat for recrystallization is conducted to inner coils. We want to avoid fracture, of course, because it requires either some means of joining the ends together again or drawing two pieces afterward — two pieces, each of which has to be pointed for each die and then drawn.

Figure 9.22
Methods for drawing seamless
tubing: (a) fixed mandrel,
(b) floating mandrel

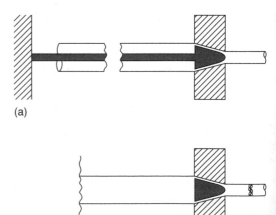

9.7 Forming Processes

Forming is a classification within secondary processing that utilizes bending, shearing, or drawing of sheet metal in most instances. However, it can also include twisting, stranding, and cabling of wire for electric and magnetic applications or for strength while retaining flexibility. An infinite number of products are formed, including automobile body parts, tools, razor blades, electrical supplies, wire rope, and electrical cable among many others.

Forming of sheet is usually done by dies, which are placed between the ram and anvil of a press. One of the most versatile forming machines is the **press brake**, which can be used to bend metals, insert preforms by crimping, and perform many other operations. Other presses, such as shear presses, are similar to the press brake in operation. They can be mechanical, hydraulic, programmable, or automated. Sheet can also be formed by rolling, an example of which is the forming of protrusions on steel reinforcing rods (*rebars*).

Wire is also formed for many applications into rectangular, square, and even hexagonal shapes. For example, rectangular wire can be wound into a tighter helix than round wire. It is drawn as a round wire, of course, but then we form it by drawing it through **Turk's head** rolls, in which the center lines for four rolls can be offset to form a square or rectangular shape. Because of flashing that is common and because of sharp corners, a final pass usually is made through a shaped die.

Round wire is also formed into cable or braid for many applications. For example, tubular **braid** is used to provide shielding from magnetic fields and steel wire is formed into **cable** to provide strength and flexibility for use in pulley systems or winding around a drum. Machines that can braid and cable were first developed for the textile industry and have been enlarged and otherwise modified to form metal braids and cables. Figure 9.23 illustrates the equipment for twisting and stranding small seven-strand cable.

We can provide an infinite variety of stranded configurations. Most steel wire ropes, however, are made with a fiber or steel core, such as the examples shown in Figure 9.24. With a steel core, for example, ¼-in. diameter wire rope has a breaking strength of more than 5500 lb.

Sheet forming uses three techniques: bending, deep drawing (which involves both drawing and stretching), and shearing. We have discussed the metal parameters that must be controlled for these processes in Section 9.1.4, but we must also control the processes themselves. For example, in **shearing** (also referred to as **stamping** or **blanking**), the shear blades that move opposite to each other deform the metal until it fractures. Both clamping pressure and blade clearance are important to control the edge configuration.

Most high-production forming is done on a press. Mechanical presses are usually quick-acting and have a short stroke, whereas hydraulic presses are slower, with longer strokes. The basic tools are the punch, which is convex, and the die, which is concave; these mate during the press stroke. Clamping is just as important in bending or drawing as it is in stamping because the sheet will frequently wrinkle if we do not provide sufficient holding pressure. The deformation during forming can lead to

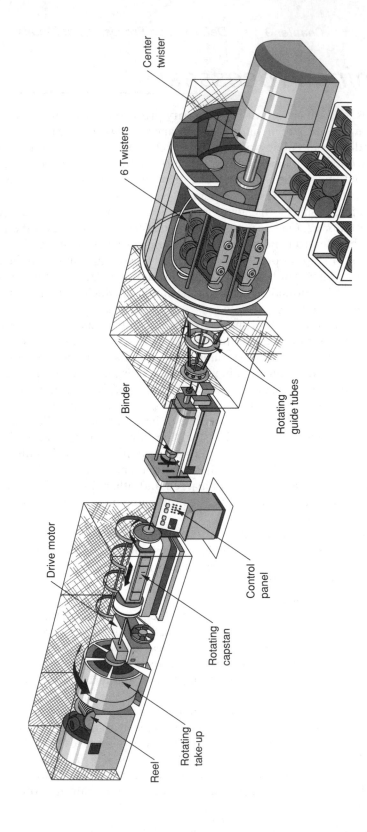

Center twister

6 Twisters

Rotating
guide tubes

Binder

Drive motor

Rotating
capstan

Control
panel

Reel

Rotating
take-up

Figure 9.23
Tandem twisting-and-stranding machine for making seven-strand cable. Six twister lines, rotating around a seventh, twist and then strand wires in a single operation. (Courtesy of AT&T.)

Figure 9.24
Cross sections of 6 × 7 class wire rope, six strands, seven wires per strand

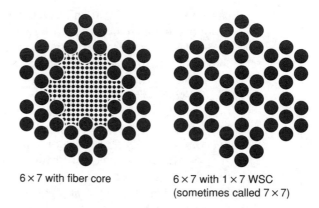

6 × 7 with fiber core

6 × 7 with 1 × 7 WSC
(sometimes called 7 × 7)

complex residual stresses and it is common practice to incorporate process anneals during production.

Complex shapes are stamped in progressive dies, where the sheet is indexed forward, and a series of forming steps are used to shape the part. Perhaps the simplest illustration of a progressive die stamping is the manufacture of a washer, shown in Figure 9.25.

Summary

Deformation processing of metals requires understanding what happens to the metal and how its properties are changed as well as understanding the processes that accomplish the deformation. Forces must be sufficient to deform the metal plastically, yet not fracture or crack the workpiece. Deformation takes place by slip

Figure 9.25
Simple progressive die forming of a washer

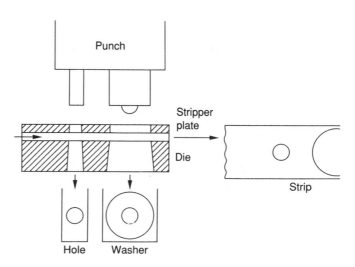

caused by resolved shear and tensile forces. The slip is facilitated by dislocations that can move through the lattice with relative ease; any prevention of dislocation motion therefore strengthens the metal. Cold working increases the number of dislocations that interact to prevent additional motion and strain (work) harden the metal. Energy is associated with dislocations, such energy also increasing with the amount of cold work. This stored energy of cold work is the driving force for recrystallization, which returns the properties to their original level. We thus can control the properties by hot-working above the recrystallization temperature or by cold-working combined with recrystallization process anneals.

Metalworking is the practical shaping of metals by deformation. Most processes use direct compression or indirect compression, particularly primary processes that break up cast ingots to form billets for further processing. Secondary processes further shape these billets into final products. Some secondary processes utilize bending and shearing forces for shaping the metal. Actual processes include forging, rolling, extrusion, wire drawing, and forming.

Terms to Remember

anisotropy	primary process
billet	process anneal
blanking	punch and die
braid	recrystallization
bull block	resolved shear stress
cable	rolling
capstan	secondary process
closed die forging	secondary recrystallization
cold working	shearing
deep drawing	shear strain
draw bench	shear stress
extrusion	slip
forging	slip system
forming	stamping
grain growth	strain hardening
Hall-Petch relation	strain ratio
hot working	temper
mandrel	texture
metalworking	true strain
penultimate grain size	true stress
press brake	Turk's head
	upset forging

Problems

1. Explain why it is more important to think in terms of true stress and true strain for metalworking rather than engineering stress and strain.
2. Explain the role of dislocations in the slip process and in strain hardening.
3. Explain the advantages and disadvantages of hot and cold working.
4. Explain texture and its importance in metalworking processes.
5. Select a product made by deformation processing and describe how it was made.
6. Make a list of four products made by each of the following:
   a. forging
   b. extrusion
   c. rolling
   d. forming
7. Explain what a cold work–anneal cycle is.
8. Describe, in your own words, two of the following processes:
   a. forging
   b. rolling
   c. extrusion
   d. wire drawing
   e. forming
9. Using the selections you made in Problem 8, explain what factors might affect productivity.
10. If a 2-ft long, 9-in. diameter billet is extruded and drawn into 0.025-in. diameter wire, how long will the wire be?
11. What diameter rod can you safely extrude in a 200-ton extrusion press if the beginning diameter is 5 in. and the extrusion constant, K, is 50,000 lb/in^2?

10

Metal-Joining Processes

Throughout this text, we have stressed the properties of metals for specific applications, selection of appropriate metals to fit those applications, and then proper processing to optimize the final product. With this in mind, we must also recognize the fundamental fact that metals are joined together in many final applications. Joining is necessary for a number of reasons:

- it is sometimes impossible to manufacture a whole product, such as carbide cutting tools or continuous electrical wiring
- many times it is more economical to produce small parts and join them together than to produce the whole, such as using mechanical threaded inserts as opposed to drilling and tapping holes
- maintenance often requires accessibility, for example, interlocked safeguards for machines and replacing parts such as bearings
- sometimes parts of a product require different characteristics than those of the predominant metal structure, such as site windows, which might be glass or plastic.

Joining actually includes a number of processes. **Mechanical joining** uses nails, screws, **rivets**, nuts, and **bolts** among other items to join similar or dissimilar materials of all types. Adhesives can also be used to join similar or dissimilar materials of all types. Metallurgical changes only occur when we use soldering, brazing, or welding processes to join metallic materials. In this chapter, we will examine all these joining methods with the assurance of product control as our goal.

10.1   Mechanical Joining _____

We can think of many applications of mechanical fastening, from nailed flooring to the cap on our toothpaste tube. It is convenient to make the distinction between removable fasteners and permanent fasteners, but in all cases we must examine the relationships of stress and properties of the fasteners as well as those of the materials being joined.

10.1.1   Riveted and Bolted Joints

In mechanical joints, there is a transfer of stress from the point at which the load is applied to the actual fastener. Figure 10.1 illustrates the stresses in a mechanical **lap joint**, where the joined materials overlap. In this figure, the stress in the bulk section is tensile, with the value equal to the load, P, divided by the cross-sectional area, $t \times w$. In the bolt or rivet, however, the stress is a shear stress, equal to the load, P, divided by the area of the rivet or bolt, $\pi d^2/4$. The rivet or bolt also bears against the short sections of the plates being joined, causing a compressive stress that we call **bearing stress**. This bearing stress is equal to the load, P, divided by the projected area of the rivet or bolt, $d \times t$. Of course, more than a single rivet is used to join large sections, with lap designs including single rows, where the spacing between rivets is termed the **pitch**, and multiple rows, where the spacing between rows is termed *back pitch*. In these cases, we find it convenient to estimate the stresses in the following way:

σ_{bulk} = strength of the gross section

σ_{net} = strength of the net section

τ = shear strength of the number of rivets $\times$ the cross-sectional area of each

$\sigma_{bearing}$ = bearing strength of the number of holes in a single row $\times$ the projection area of each

Figure 10.1
Stress in a bolted or
riveted joint

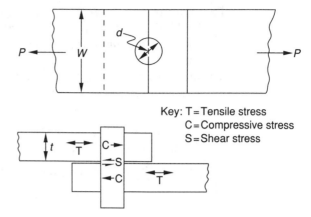

Key: T=Tensile stress
C=Compressive stress
S=Shear stress

Sample Problem 10.1

Two aluminum plates, each ¼ in. thick and 10 in. wide, are joined by seven ½-in. diameter aluminum alloy rivets equally spaced in a single row. The joint transmits a tensile force of 10,000 lb. Find the unit tensile stress in the bulk plate and in the net section, the shear stress in the rivets, and the bearing stress between the rivets and the plates.

Solution

$$\sigma_{bulk} = \frac{10,000}{10 \times \frac{1}{4}} = 4000 \text{ lb/in.}^2$$

$$\sigma_{net} = \frac{10,000}{(10 - \frac{7}{2}) \times \frac{1}{4}} = 6150 \text{ lb/in.}^2$$

$$\tau = \frac{10,000}{7[\pi(\frac{1}{2})^2 / 4]} = 7275 \text{ lb/in.}^2$$

$$\sigma_{bearing} = \frac{10,000}{7 \times \frac{1}{2} \times \frac{1}{4}} = 10,428 \text{ lb/in.}^2$$

Failure can occur in any of the modes, shown in Figure 10.2, if the ultimate strength of the metal is exceeded. Of course, it is much easier for repairs to be made if the rivets rather than the parts being joined fail, so we should design joints and make material selections accordingly. Another possible mode of failure for bolts and removable fasteners (Section 10.1.2) is stripping of the threads by the clamping load.

The bearing stress borne by the engaged threads is equal to the clamping load divided by the contact area, which is $\pi(d^2 - d_r^2)/4$, where d is the major thread diameter and d_r is the minor thread diameter. Threads do not take an equal share of the load but do transfer some load to other threads when they deform. Conservative designers assume all the load is supported by one thread. The rules of thumb for reasonable length of thread engagement are

- length equal to outside thread diameter (steel)
- length equal to one and one-half times outside thread diameter (brass)
- length equal to twice the outside thread diameter (aluminum, plastic)

Figure 10.2
Failure modes for a riveted joint: (a) failure in tension, (b) bearing failure, (c) shear failure

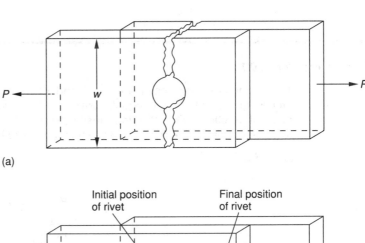

(a)

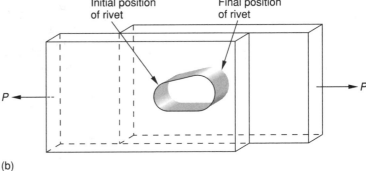

(b)

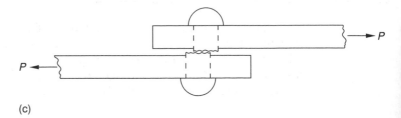

(c)

10.1.2   Removable Fasteners

We can consider screws and nuts and bolts as removable fasteners, but there are many other types of removable fasteners as well. There are clamps, such as hose clamps, clips and harnesses for electrical equipment, and many more. But the purpose of this text is to examine the metallurgical aspects of fasteners rather than their variety of types. Let's look at a fastener problem that demonstrates the importance of knowing about metals, their processing, and their application.

Case Study 10.1

Assembly Using a Stickscrew

Stickscrews make rapid assembly of many electronic and mechanical devices possible. Using a screw gun with a hexagonal adaptor, joining can occur almost continuously because the **stickscrew** breaks at the undercut diameter at a predetermined design torque. Figure 10.3 shows the design for a case-hardened steel designed to break away at a torque of 6–8 in.-lb. These were self-threading screws used in conjunction with zinc die castings. Although the assembly process is cost efficient, manufacturing costs skyrocketed when a series of stickscrews broke in the threads away from the undercut diameter. Units had to be removed from the assembly line and completed by hand after the broken screws were removed.

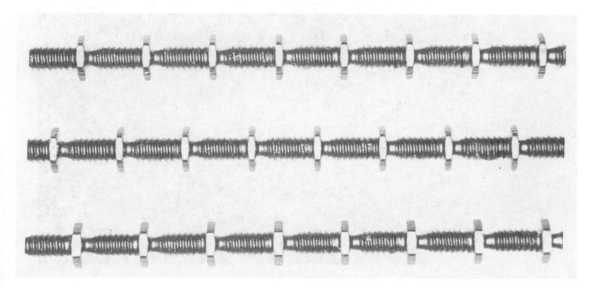

Figure 10.3
Stickscrews used for semiautomatic assembly

Stickscrews were torque tested and samples that broke away at areas other than the undercut section were found to have a higher breaking torque, 9–10 in.-lb. The metallurgical history included roll-formed threads, followed by a carburizing surface-hardening treatment. Metallographic examination showed the case depth was almost twice as much for the stickscrews that did not meet torque requirements, as shown in Figure 10.4, which indicated that the undercut area was substantially stronger than needed. Scanning electron microscopy showed significant stress concentration of the roll formed threads which led to premature failure (see Figure 10.5). The production problem was corrected by altering the case-hardening specifications to include case depth and incoming torque testing of all new stickscrew lots.

There are many other types of metal fasteners used for mechanical joining, including staples, snap-in plastic and metal fasteners, preformed inserts that are crimped in place, press- or force-fit joints, and shrink-fit joints that depend on thermal expansion differences.

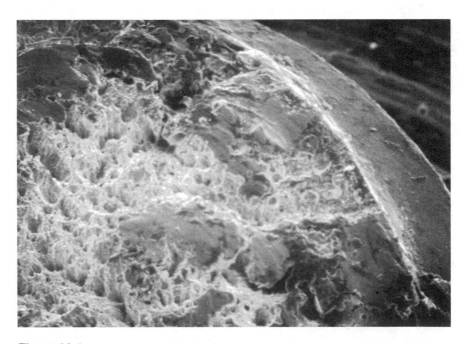

Figure 10.4
SEM micrograph of the stickscrew fracture initiated at a defect in the thread root, away from the designed separation (100×)

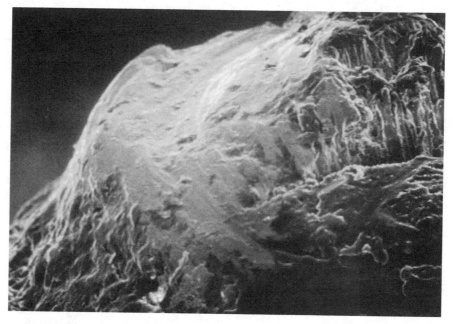

Figure 10.5
SEM micrograph showing the case-hardened rolled thread near the fracture
surface of Figure 10.4 (151×)

10.2   Adhesive Bonding

Adhesive joints are the first nonmechanical permanent or semipermanent bonds that we will study. The adhesive bond is actually a chemical surface bond classified as a *cohesive bond*. Adhesive bonding has been used for many years in the packaging, furniture, publishing, and footwear industries. The number of adhesives and their applications grows every year, and currently they are common in electronics, aerospace, building construction, automotive, and many other industries. Perhaps because of such extensive usage, we must stay with our fundamental approach to materials — to design the joint, select the correct adhesive material, and fabricate the joint properly.

In designing the joints, we must balance the many advantages of adhesives with their limitations or disadvantages for each application. These designs consider the types of forces on the joints, including peeling forces as well as tensile, compressive, and shear forces. Table 10.1 lists some of the more important advantages and disadvantages of adhesives for industrial applications.

Table 10.1
Advantages and disadvantages of adhesive joining

Advantages
1. Stress distribution is possible over a large area
2. Can join dissimilar materials
3. Can join fragile and/or thin materials
4. Joins materials of complex shapes
5. Adhesive dampens vibrations and seals joint
6. Bonding is a low temperature process

Disadvantages
1. Surface preparation and cleanliness are critical
2. Careful selection from the large number of adhesives may be difficult
3. Joints cannot be readily disassembled
4. May require long curing times under heat and pressure
5. Upper service temperature is limited
6. Fixtures or jigs for bonding are often needed

Designers must carefully weigh these factors to plan the most effective and economical joining process. The type of material being bonded is known as the **adherend** and can be any number of metals, ceramics, or plastics. Matters such as cleaning procedures and strength of the joint can be affected by the adherend. All service conditions must be evaluated and all procedures for assembly and bonding must be evaluated beforehand.

Surface preparation is essential for adhesive bonding. There are two general procedures that we can use, chemical and/or solvent cleaning and abrasive cleaning. Both of these remove adherend material, leaving a clean surface that promotes adhesion. In some cases, we can also have mechanical bonding by interlocking the surfaces at a microscale, such as in wood-to-wood and paper-to-paper bonds, but in metal-to-metal, ceramic-to-ceramic, and plastic-to-plastic bonds, only the chemical or cohesive bond provides the adhesion. Some metals require both abrasive cleaning and acid etching to promote adhesion. Adhesives can be applied with rollers, by brush, by trowel, and even by extrusion coating. The most uniform coating is achieved by rolling, either manually or between pressure rolls.

10.2.1 Types of Adhesives

Adhesives are polymeric, that is, they are organic, long-chain, high molecular weight materials. These polymers or plastics are made up of thousands of repeating chemical units called *mers* that are joined together. Bonds along the chain are strong cohesive bonds but bonds between chains are weak secondary bonds. Thermoplastics, such as polyethylene, soften and are deformable as they are heated and are capable of being reshaped. Thermosetting plastics, however, can have strong bonds between the chains (cross-linking) that are not reversible once formed.

There are many adhesives available and new ones are continually appearing, but we can classify them as three main types: chemically reactive, evaporation or diffusion type, and hot melts. **Chemically reactive** adhesives include two-part epoxies, phenolics, and polyurethanes that chemically react (polymerize) to form strong bonds. **Evaporative adhesives**, such as acrylics, natural or synthetic rubbers, and vinyl resins ("white glue") are dissolved in solvents that evaporate, leaving a strong bond. **Hot melts**, thermoplastic materials such as polyethylene, require properly formulated resins and special application equipment to promote bonding. Table 10.2 summarizes some of the major adhesives and their applications.

10.3   Metal Joining

The most common metal-joining methods are soldering, brazing, and welding; in these cases, the adherend and joining materials are metallic and a **filler** metal is added during the joining process. The methods can be differentiated because **soldering** and **brazing** utilize filler metals that have very different compositions from the adherend metal, and the adherend metal is not melted during joining. **Welding** uses filler metal compositions similar to the adherend metal, and the adherend metal is melted during joining. Brazing and soldering can be further distinguished by the melting temperatures of the filler metal alloys. Brazing takes place at temperatures in excess of about 800°F, whereas soldering is done at temperatures lower than this. In all these joining methods, we must consider joint design, adherend composition, filler metal composition, **cleanliness**, and heating methods.

10.3.1   Joint Design

We must be concerned with how we design joints because they must meet all the strength and other property criteria that are required of the adherends. The common types of joints used for soldering and brazing are shown in Figure 10.6; those for welding are shown in Figure 10.7. Joints are inherently weaker than the adherend because they are cast metals subject to the limitations of solidification. **Butt joints**, where parts are joined end to end, have limitations, therefore, and larger joint areas are preferred. Such is the advantage of scarf joints where the area is increased by angling the joint through its thickness. The ability of lap joints to withstand stress depends on the length of the lap area. A rule of thumb is that the length of the lap should be three times the thickness of the thinner adherend metal.

Table 10.2
Some common industrial adhesives

Classification	Type	Characteristics	Common forms available	Product usage
Chemically reactive	Acrylic	Thermoplastic, polymerizes with pressure, must be ventilated, flexible to head compositions, uses solvents	One or two component liquids	Composites, electronics, automotive, consumer
	Formaldehydes (phenolic) (resorcinol)	Thermosetting, heat and pressure required, waterproof bonds	Film, liquid, powder	Plywood, sandpaper, grinding wheels
	Epoxy	Thermosetting, excellent bond to every material except polyethylene, silicone, and fluorocarbons, chemically resistant	Two-part pastes, dispersion films	Sandwich composites, printed circuit boards
	Polyurethanes	Thermosetting, flexible, good strength at cryogenic temperatures	Dispersion liquid, powder, film	Cryogenic sealants, metal-to-plastic bonding

It is common practice to use solder or braze **preforms** (material that has been pre-shaped) for many applications, but some, such as the grooved joint of Figure 10.6, actually require machining or forming of the adherend metal. In weld joints, the weld penetration is critical. Fillet welds are used for the joints depicted in Figure 10.7, but the use of beveling, such as in the single V butt, promotes thorough penetration. It is common to employ a safety factor to compensate for voids and/or stress concentration factors caused during the actual brazing or soldering operation.

10.3.2 Soldered and Brazed Joints

When adherends are joined by soldering or brazing, microalloying occurs at the joint. How easily this takes place depends on the type of adherend, cleanliness, and the composition and temperature of the solder or braze alloy. The filler metal

Classification	Type	Characteristics	Common forms available	Product usage
Evaporation	Silicones	Thermosetting, pressure or heat curing adhesive or sealant, high cost		Pressure-sensitive tapes, sealants
	Rubbers natural butyl nitrile neoprene styrene butadiene	Thermoplastic, water soluble	Liquid	Tire cord, dry-seal pressure-sensitive tapes, contact cements
	Vinyl resins	Thermoplastic, water soluble, basis for common household "white glues"	Liquid, powder, dispersion	Furniture, pack-aging
	Cellulosic	Thermoplastic, general-purpose	Liquid, solid, powder	Household cements
Hot melt	Polyethylene, polystyrene, polyvinyl acetate	Thermoplastic, quick setting, diffi-cult to make large joints, "glue" gun necessary	Solid	Leather, book-binding

spreads, or wicks, because of capillary forces when the wetting characteristics of the adherend metal are correct. In some cases, cleaning (discussed under "Adherend Surface Cleanliness") and fluxing is all that is necessary. For example, copper is easy to solder to when cleaned and fluxed properly. Figure 10.8 illustrates the microalloy-ing in a solder joint between copper sheet, using a bismuth-base solder containing indium. The X-ray mapping in the scanning electron micrographs shows the diffu-sion of copper into the joint and diffusion of indium into the copper.

The strength of the joint depends not only on the microalloying, but on the clearance at the time when wicking takes place. **Clearance** controls the thickness of the joint itself. Figure 10.9 illustrates the effect of joint thickness for stainless steel brazed with a silver-base braze alloy; maximum strength of the joint corresponds to joint thickness of 0.0015 in. Good joint strength is achievable with a simple slip-fit if the same metals are being joined, but caution must be taken when dissimilar metals are being joined because of differential thermal expansion between room tempera-ture and the soldering or brazing temperature.

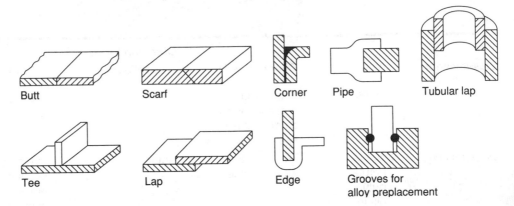

Figure 10.6
Common braze joint designs

Stainless steel is actually quite difficult to join because it is hard to clean adequately. Special techniques, such as electroplating with nickel, are frequently resorted to for production. Figure 10.10 shows the brazed joint between sulfamate nickel-plated 1215 carbon steel and similarly plated 303 stainless steel. The dark phase of the braze joint is nickel containing copper and zinc, whereas the light phase is silver. The braze alloy is silver-base containing copper, zinc, and a small amount of nickel.

Adherend Surface Cleanliness

Adherends must be clean, that is, free of oxidation, grease, and so on, in order to form a reliable joint, whether by soldering, brazing, or welding. In addition, they must remain so up to the joining temperature. We start with cleaning the surfaces,

Figure 10.7
Common weld joint designs

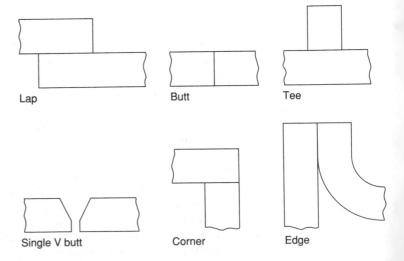

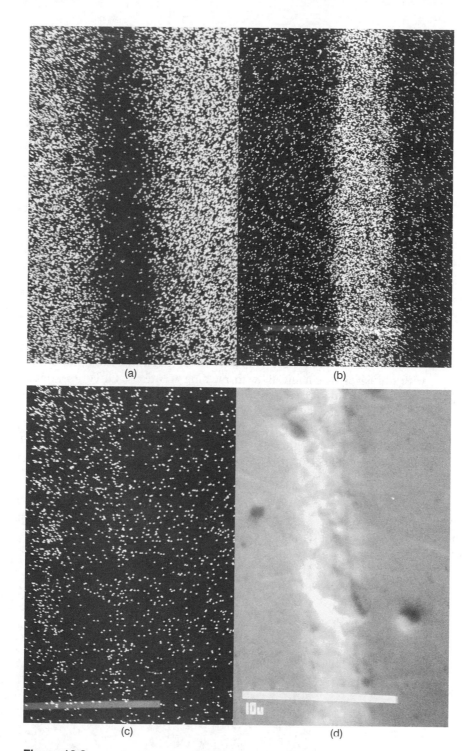

(a)

(b)

(c)

10u

(d)

Figure 10.8
SEM X-ray mapping of a soldered copper joint using a bismuth-base solder
containing indium: (a) copper X-ray map, (b) bismuth X-ray map, (c) indium
X-ray map, (d) SEM micrograph of the solder joint (3400×)

Figure 10.9
Effect of joint thickness of stainless steels brazed with a silver-base braze alloy (Courtesy of Handy and Harman.)

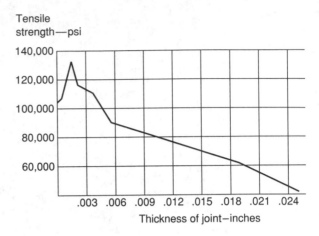

first by removing grease and oil with an environmentally safe solvent or by alkaline cleaning. If the metals are oxidized, we can remove the scale either mechanically with abrasives or chemically with pickling solutions. Of course, once the surface is clean, we should join as soon as possible in order to prevent reoxidation.

Flux is a chemical compound that we use to keep the surface clean as the metals are heated to the joining temperature. This obviously means that the fluxes we use in brazing or welding have to withstand higher temperatures than those we might use for soldering. Fluxes are also surfactants that promote wetting at the join-

Figure 10.10
SEM micrograph of a shear joint between 1215 carbon steel and 303 stainless steel (1000×)

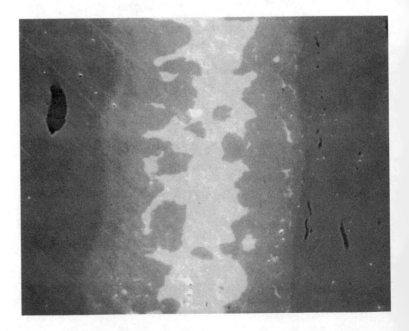

ing temperature. They are available in solid (fused powders), liquid, and paste forms for various applications. Because of their extreme corrosiveness, fluxes must be removed after joining, a problem particularly in electronics, where gas-phase fluxing, elimination of ozone-depleting chlorofluorocarbons, and no-clean fluxes are popular issues.

Solder and Braze Compositions

Solder and braze alloys are selected on the basis of melting or soldering temperatures, mechanical properties (usually strength), electrical characteristics, economy, and, of course, safety. Melting temperatures are important to the application, such as fusible links for a sprinkler system, but soldering temperatures are important for the processing.

In all cases, the alloys are based on eutectic alloys that are compatible with the metals being joined, that is, microalloying at the adherend surface occurs. The actual compositions are near the eutectic composition because the solidification shrinkage of thin sections where the two-phase temperature range is extensive can cause severe stress conditions for the joints and inferior joint bonds. For this reason, many alloy compositions have multiple alloy additions in order to tailor the eutectic composition and thereby limit the melting-temperature range. In order to promote wetting, the actual alloy temperature is higher than the liquidus, typically about 75–100°F.

Perhaps the most common solder alloys are the lead-tin alloys, which are available in most hardware stores for home use and are the mainstay of the electronics industry. These are used to join the copper conductor to components on printed circuit boards and as corrosion protection for the underlying printed wiring during processing. Figure 10.11 shows the Pb-Sn phase diagram and how the alloy strength and conductivity is altered by composition. The 60-40 alloy, near the eutectic composition, is the standard for electronics because of the unique combination of strength, conductivity, and melting temperature.

Soldering and brazing do not differ, except in the melting temperature range; solders melt at temperatures below 800°F whereas braze alloys melt at higher temperatures. There are many alloy compositions, but the more common ones, along with their melting ranges and applications, are shown in Figure 10.12.

Producing Solder and Braze Joints

In addition to the cleaning and flux practice and the choice of solder or braze composition, the methods for producing the joints are of extreme importance for total quality management and productivity. There are three basic processes that we use for making solder or braze joints: manual, batch, and **continuous processes**. We also can distinguish the various processes by the energy used to heat the joints to the joining temperature or to promote wetting. For example, we are all familiar with electrically heated guns or propane torches for soldering, but we could also make solder joints with difficult metals such as stainless steels by combining ultrasonics with a hot plate; these are all **manual processes**. In brazing, it is common practice to use braze preforms and induction heating of the area where the joint is to be made.

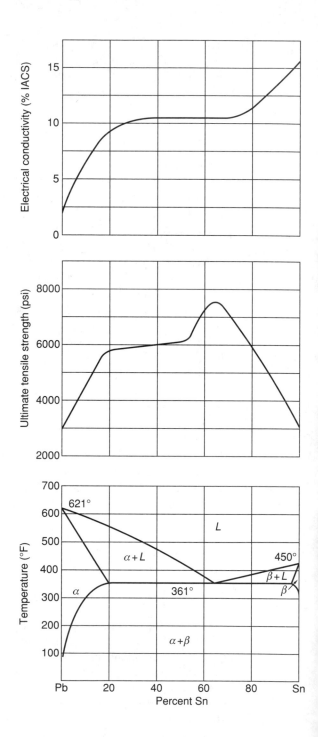

Figure 10.11
The Pb-Sn phase diagram and influence of composition on strength and conductivity of Pb-Sn alloys

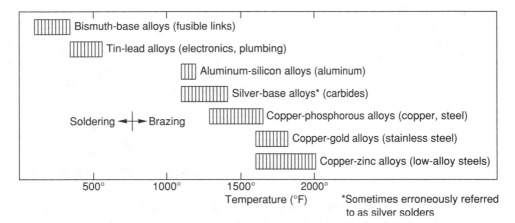

Figure 10.12
Melting ranges and applications (materials joined) of some common solder and braze alloys

When multiple joints are formed at the same time, fixturing (holding in place) is vital to preserve the shape of the joined parts. Batch production by dip soldering or **dip brazing**, dipping parts in a bath of braze alloy, is very dependent on fixturing practice. When the higher temperatures of brazing are considered, another factor becomes important. The stresses of cooling from the brazing temperature to room temperature for different metal thicknesses can cause distortions of the parts; in these cases, we must stress-relieve the parts before fixtures are removed and, in extreme cases, must reshape or deform the parts prior to stress relieving to ensure product quality. Furnace soldering and brazing are also common **batch processes** for joining.

For high production rates, we find that the most common soldering method is a conveyorized wave-soldering method. The melting point of the solder is low enough that we can design methods to control the liquid phase. For brazing, however, the melting temperature is too high to permit such control and we find the most common production method a conveyorized furnace with controlled atmosphere. For example, the stainless steel–carbon steel braze of Figure 10.10 was made in a conveyor furnace with a hydrogen atmosphere.

Of the automated soldering methods, **wave soldering** is predominant, particularly for the electronics industry. When we solder components to printed circuit boards, the components must not be exposed to the soldering temperatures for any length of time, yet the board must remain in contact with the solder long enough to ensure a sound bond. The action of the wave is such that we must have maximum contact time with the wave and, at the same time, minimum time when exposed to the atmosphere where the solder will oxidize. Oxidation limits the life expectancy for the specific type of pumping system. The most common wave-soldering systems have a conveyor onto which the boards are placed. They travel past a flux applicator, which might use brushing, spraying, rolling, or foaming to apply the flux, then past a

Figure 10.13
Schematic drawing of the
wave-soldering process for
printed circuit manufacture

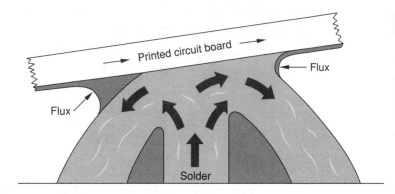

preheating system, then over a solder wave, such as depicted in Figure 10.13. Solder waves can be unidirectional, such as solder jets, or bidirectional, which is preferred to minimize "icicling," a cosmetic defect. The conveyor is usually inclined about 4° as well in order to minimize such a defect.

An automated soldering system is not complete if it does not contain a cleaning system to remove residual flux, which can be corrosive. Although this problem is most evident in the electronics industry, corrosion caused by lack of flux removal has caused many problems in plumbing and other soldering applications. Figure 10.14 illustrates one possible aqueous alkaline cleaning procedure.

Solder and Braze Defects

Solder and braze defects can be caused by many factors, such as contamination or moisture. They can take the form of inadequate or excess solder or braze, **dewetting** (lack of adherence of solder to metal), voids or blowholes, deposits, and many others. The source of the problem can be in design, the adherend materials, the flux application process, the solder or braze process, or the cleaning practice. Figure 10.15 illustrates some common defects encountered when soldering components to printed circuit boards. In this particular case, bridging and incomplete filleting were corrected by redesign of the joint or reorientation of the component, whereas the dewetting and nonwetting were corrected by surface cleaning and proper flux practice.

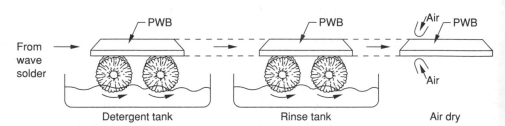

Figure 10.14
Schematic drawing of a flux-removal procedure for printed circuit manufacture

Sometimes we are unable to adequately feed the liquid solder or braze into a joint. In Figure 10.16, a void was created by solidification shrinkage of the silver-base braze at a right angular junction where the joint is in tension at the bottom and in shear to the side. In this case, the void was not a problem because of the adequate contact area.

Case Study 10.2

The Solder Volcano

PC Warehouse is a fabricator of printed circuit boards and performs custom assembly and joining for small computer manufacturers. Their production is limited to single- or double-sided boards and does not include multilayer boards. Processing includes CNC drilling, cleaning, and electroless plating of the holes where component leads are to be inserted. After a water rinse, the boards are stored for two to four weeks before assembly and wave soldering.

Initially, drilled boards were removed from storage shelves, components were automatically inserted, and the boards were placed on the wave-soldering machine. Early pro-

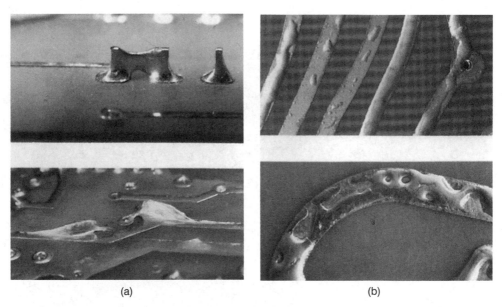

(a)   (b)

Figure 10.15
Common solder defects in electronic applications: (a) solder bridge (short circuit) between component leads (*top*) and film or webbing between circuitry (*bottom*), (b) dewetting (*top*) and nonwetting (*bottom*)
(Courtesy of AT&T.)

Figure 10.16
SEM micrograph of braze between 1215 carbon steel and 303 stainless steel diaphragm (125×)

Figure 10.17
SEM micrograph of a blow-hole in a solder joint of a printed circuit through the hole (75×)

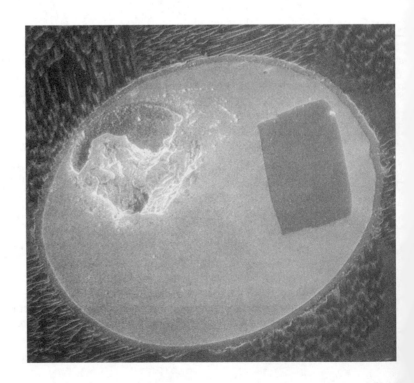

duction runs were successful, but after several months, blowholes (humorously referred to as volcano eruptions) were increasingly encountered, such as that shown in the solder cross section of Figure 10.17.

After investigation, it was found that the drilled boards had absorbed moisture during storage in the hot, humid summer months. Moisture vaporized during the soldering, causing the blowholes as the vapor escaped. Correction was easily accomplished by incorporating a bake-out heat treatment for all boards prior to assembly.

10.3.3 Welding

Welding differs from other metal-joining processes because, in most instances, the adherend metals are melted. Weld joints, therefore, usually have all the characteristics of a cast metal. The metal area near the weld also is heated to near the melting point and its properties can be affected by the welding as well. We call this area the **heat affected zone**, or HAZ. There are many welding processes, but the most popular are arc welding, resistance welding, and gas welding. We will examine these in detail and look at some of the more recently developed processes such as electron beam welding and laser welding. Table 10.3 lists the welding processes we normally encounter.

Welding Processes That Involve Deformation

There are many common welding processes that are very useful in metal fabrication, but we do not necessarily think of them as welding processes. For example, the coinage minted in the United States is a clad metal, which is welded by rolling the bimetal sheet in a rolling mill. The process can be done either hot or cold, but is very dependent on surface cleanliness prior to bonding. Multifilamentary superconductors containing filaments of superconducting alloys are made by stacking rods of the alloy in cleaned copper tubes with round internal diameters and hexagonal outside diameters. These are then inserted into a copper casing, evacuated, and extruded to form the solid composite.

Table 10.3
Welding processes

Chemical	Electrical	Mechanical	Other
Oxyacetylene	Shielded metal arc	Pressure (cold)	Electron beam
Other gas	Shielded tungsten arc	Explosive	Laser
Thermit	Submerged arc		Thermocompression
	Resistance		

These types of welding require pressure when the materials are constrained, but other types of cold welding can be accomplished only with pressure. For example, copper wires broken in wire drawing can be rejoined by forcing cut ends together with only pressure, saving production time.

Thermocompression welding, which uses both temperature and pressure, has been developed for joining 0.001-in. diameter gold wires to metal lead frames and silicon chips in microelectronic devices; the technique has also found applications in joining to thin films. The most popular versions are the wedge bond and ball bond, shown in Figure 10.18. In the wedge bond, the wire is deformed by a wedge placed against the surface to be bonded to, and either the wedge or the surface is heated. In the ball bond, the wire is continuously fed to the joint and a microjet of hydrogen gas melts it at the surface as pressure is applied.

The last method that requires mechanical pressure is explosive bonding. In this method, an explosive detonation forces two materials in close contact together, forming a wavy interfacial bond. The technique is extremely limited because it is inherently dangerous, requiring personnel with expertise in explosives.

Chemical Welding Processes

The two basic chemical welding processes are Thermit welding and oxyfuel **gas welding**. Acetylene (C_2H_2) is the most popular fuel gas, but others such as MAPP gas (methylacetylene propadiene) are also common. **Oxyacetylene welding** was the first to be developed as a manufacturing process, made possible by the development of practical pressurized torches that combine the acetylene and oxygen in the right proportions. Almost all of the oxyfuel welding involves melting of the metals being joined; filler metals are needed because of the gap between the pieces.

Figure 10.18
Thermocompression welding configurations

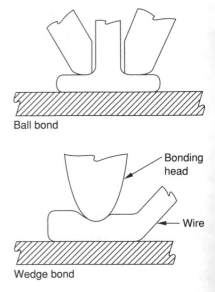

Ball bond

Bonding head

Wire

Wedge bond

The combustion of acetylene and oxygen at the tip of the torch produces temperatures of about 6000°F (3000°C) in a two-stage reaction. With proportional amounts of oxygen and acetylene supplied from tanks to the torch nozzle, the reaction that takes place is

$$C_2H_2 + O_2 \rightarrow 2CO + H_2$$

These products react with oxygen from the air:

$$2CO + H_2 + 1.5O_2 \rightarrow 2CO_2 + H_2O$$

Safety dictates that oxyfuel welding be conducted in well-ventilated areas because of the depletion of atmospheric oxygen in the process; 60% of the oxygen in the flame comes from the atmosphere. By controlling the proportions of acetylene and oxygen to the torch, neutral, oxidizing, or reducing (carburizing) flames can result. Neutral flames are used in welding to prevent carbon or oxygen contamination. We use the oxidizing flame for metal cutting and sometimes for brazing operations. In metal cutting, special torches are required that add more oxygen because the metal is actually oxidized, not melted during cutting.

Oxyfuel welding is applicable for ferrous and nonferrous alloys but has been largely replaced by other welding processes, except for repair work and a few special applications. Its use for metal cutting, however, is commonplace, even encompassing computer numerically controlled and robotic operations. There are numerous applications as well for heating metals in manual forming operations.

Thermit welding is the only other chemical welding process that is used to any extent. It is practiced only at remote sites inaccessible to other welding processes. The chemical reaction is exothermic, producing temperatures of the same order of magnitude as oxyfuel processes in very short times. The basic reaction is

$$3Fe_3O_4 + 8Al \rightarrow 9Fe + 4Al_2O_3$$

Although the process is not explosive, expertise is required and safety precautions are necessary.

10.3.4   Electrical Welding Processes

Welding processes that derive their heat from electricity include arc welding, which is the most popular industrial welding process, and resistance welding, which includes spot welding, seam welding, and projection welding.

Arc Welding

Arc welding uses consumable or nonconsumable **electrodes** to form an electrical circuit such as that depicted in Figure 10.19. Both DC and AC circuits are used. In DC circuits, straight polarity is defined as the condition when the workpiece is the anode (positive) and the electrode is the cathode (negative). Reverse polarity refers to a negative workpiece and positive electrode. Because more heat is generated at the anode, straight polarity is recommended for easy-to-weld metals, and reverse polarity is recommended for hard-to-weld metals and for vertical or other out-of-position welds where more control is necessary. Of course, polarity is alternated when AC power is used.

In consumable electrode welding, the electrode is melted and becomes filler metal in the weld area. When nonconsumable tungsten electrodes are used, filler metals must be fed to the weld area separately. As in soldering and brazing, we need to keep the adherend metals clean when heating up to the welding temperature. This is accomplished by using an inert gas or solid flux to shield the arc. Thus we use the terminology gas-metal arc welding (**GMAW**), gas-tungsten arc welding (**GTAW**,) and shielded-metal arc welding (**SMAW**). Helium and argon are used as the inert gases.

Bare electrodes are used only for automated welding and then only with the inert gas feed and automatic devices that maintain the arc length. For manual welding, flux-coated electrodes are used. These coatings serve a number of desirable functions:

- they provide an inert atmosphere
- they stabilize the arc
- they serve as flux
- they reduce spatter
- they provide slag to entrap impurities

Coated electrodes are classified on the basis of tensile strength of the filler metal, the welding position, and the type of coating. For example, an E6016 electrode would have deposited weld metal of 60,000 lb/in.^{2} tensile strength, be usable in all weld positions (flat, horizontal, and vertical), and have a low-hydrogen coating that

Figure 10.19
Schematic diagram of arc-welding circuit

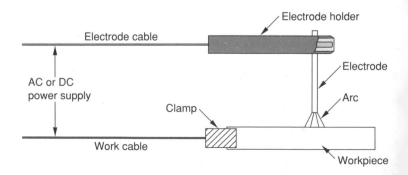

limits any dissolved hydrogen in the weld metal, thus eliminating microcracking. All electrodes have standard color codes.

Most manual arc welding is SMAW, using the coated electrodes, and most automatic arc welding is GMAW. Figure 10.20 illustrates these processes; in both processes, a weld bead is laid with the electrode at an angle to the workpiece of 60° to 80°. This figure also illustrates the important parameters of a weld — the depth and heat-affected zone. Note that in SMAW, the flux leaves a slag layer on the weld bead that protects it from oxidation during cooling. This layer is easily chipped away afterward. The heat from the arc represents energy that not only melts both the metals being joined and the filler metal, but also transfers the molten metal in the bead. Although this arc is complicated, its penetration is proportional to the current; this is particularly important when welding metals of dissimilar thickness. Current needed to melt the thicker component can be excessive for the thinner component, effectively blowing it away. Consequently, the final weld does not necessarily comply with the design and can cause premature or unplanned failures. In any case, penetration in a single pass is limited to less than ¼ in. for steel.

Figure 10.20
Schematic diagrams for manual (SMAW) and automatic (GTAW) arc welding: (a) shielded-metal arc welding (SMAW), (b) gas-metal arc welding (GMAW)

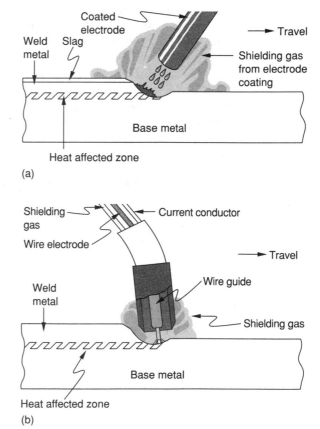

Resistance Welding

Resistance welding is used extensively in many industries, such as automotive and electronics, for both spot welding and seam welding applications. Heat for resistance welding is generated by passing a high current at low voltage through the workpieces to be joined. The heat, H, that results is given by the equation

$$H = I^2Rt$$

where R is the resistance of the circuit, I is the current, and t is the time electricity flows. Alternating current is used in most instances.

The total resistance is equal to the sum of the resistances of the workpieces, the contact resistance between the electrodes and the workpieces, and the resistance at the workpiece interfaces where the weld is to be made. It follows that good conductors such as aluminum or copper are harder to weld in this way and it also follows that the highest resistance should be the contact resistance between the workpieces so that the highest temperature is reached where the weld is to be made. Pressure applied by the electrodes lowers the contact resistance and also contributes to pressure welding. It is necessary, therefore, to control both the magnitude and time of duration of the electrode pressure as well as the contact area and the cleanliness of the workpiece surfaces.

We need not be limited to spot welding; we can substitute a wheel for the electrodes and maintain the current as the wheel moves. In this fashion, we make a **seam weld**. We can also alter the weld patterns of larger pieces by creating bumps or projections where contact is made. Passing a current then welds the pieces where the projection contacts are made, thus forming a **projection weld**. These methods, depicted in Figure 10.21, are all used extensively.

Welding Processes Using Newer Technologies

There are several welding techniques that use newer technologies to provide the heat required for welding. Among these methods are electron beam welding, laser welding, and plasma welding. **Plasma welding** is similar to gas welding with the

Figure 10.21
Examples of resistance welding procedures

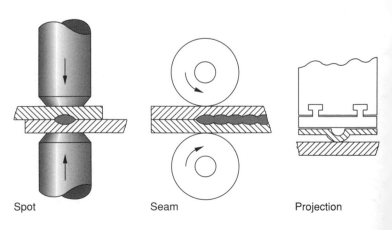

Spot   Seam   Projection

exception that the ionized gas or plasma is much hotter than oxyfuel gases. The energy sources in **laser welding** and **electron beam welding** are the laser, a high-energy beam of coherent light, and electrons given off by a heated tungsten filament. In each case, the beams can be focused or concentrated by magnetic or optical lenses that provide deep, narrow penetration or diffuse shallow welds.

Electron beam welding must be conducted in a vacuum; the better the vacuum, the deeper the penetration possible and the better depth-to-width profile. No filler metal is needed and no shielding is required because of the vacuum. Laser welding is similar to electron beam welding, with two important advantageous distinctions. Laser welding does not require vacuum conditions and it is easily automated because of the optical focusing.

10.3.5 Welding Metallurgy

We are interested in the joining of metals because of the important goal of achieving desired properties. In welding, this goal is critical because of the melting, resolidification, and heat-affected zone in the weld area. For these reasons, we define weldability of a metal as the capability of achieving the desired properties and characteristics for specific applications. We must consider the welding process, temperature reached in joining and thermal stresses, segregation, shrinkage, heat-treating characteristics, and even other parameters such as corrosion potential.

Mild or low-carbon steels are probably the most common class of metals we as manufacturing personnel deal with. If we limit discussion to arc welding, weldability is excellent for these steels but can become complicated for higher carbon and alloy steels. For example, if we use SMAW with E6012 electrodes and tight-fitting joints, the only problems we should encounter are slag entrapment, shrinkage, porosity, and perhaps cracking in the case of deep penetration welds. The microstructure is affected, with the fusion, or melting, zone being a cast microstructure, but the heat-affected zone changes from the base microstructure through a coarse recrystallization nearest the fusion zone. For thick sections, beveled edges are used in combination with multiple passes. Figure 10.22 depicts the weld macrostructure for a beveled butt weld made in mild steel pipe.

Medium carbon steels (0.25–0.50% C) present a different picture because of the potential for martensite formation. Low-hydrogen electrodes must be used and post- and/or preheating of the weld area is the practice. Preheating minimizes any martensite formation, whereas postheating tempers any martensite that has been formed.

High-carbon steels, containing more than 0.50% C, definitely call for pre- and postheating and are always test welded beforehand. Section thickness plays a role in this requirement for testing. For example, in welding 1065 steel with an E7024 electrode with no pre- or postheating, no cracking occurred and $\frac{1}{16}$ in. of martensite formed, but when the same welding was attempted for sections twice as thick, the thickness of the martensite increased and cracking was observed. Austenitic stainless steel electrodes also can be used to prevent any cracking and provide notch toughness.

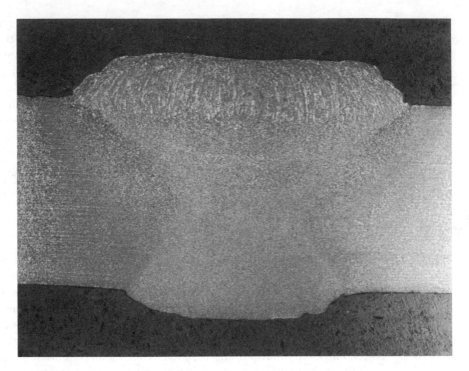

Figure 10.22
Macrostructure of beveled butt welds joining mild steel pipe (7×)

Alloy steels are more difficult to weld and require pre- and postheating plus multiple passes, using higher strength electrodes than those required for plain carbon steels. Austenitic stainless steels present a unique problem in welding because of their susceptibility to intergranular corrosion (see Chapter 11). This is preventable by using low-hydrogen stainless steel electrodes.

10.3.6 Welding Defects

Although there can be a number of defects that affect the strength or other properties of a welded joint, the most common problems are porosity, incomplete penetration, and cracking. Defects less frequent in number include microstructural problems, slag inclusions, and others. Porosity, of course, reduces the apparent strength because of stress concentration. It can be caused by gas absorption in the molten weld and subsequent release during solidification, a common problem encountered in welding aluminum. Incomplete penetration can be caused by insufficient heat to melt the adherend metal. This can be a problem when joining metals of differing thickness or by not preheating a joint and not using multiple passes. Cracking is the result of such phenomena as thermal stresses during cooling, shrinkage, martensite formation, and deep, narrow welds. Case Studies 10.3–10.7 present situations where weld defects have caused failures.

Case Study 10.3

Gas Explosion

A 6-in. steel pipeline installed 30 years ago cracked in cold weather, leaking gas that caused an explosion when ignited. The crack occurred along a weld that was arc welded, with a butt joint where ends had not been beveled. Arc welding was state-of-the-art practice but did exhibit incomplete penetration for the first pass where there was no preheating. As a result, there was sufficient stress concentration from the thermal shift to cause failure in the cold weather.

Case Study 10.4

Failure to Make the Curve

Jack is an independent truck driver in Maine. He purchased a new trailer for his tractor and was carrying his first full load, Maine potatoes for delivery in New York. When he entered a curve to the left on the highway while driving at the speed limit, he felt a load shift and lost control. The tractor and trailer overturned, coming to rest only a few feet from the road. Accident reconstruction revealed that many welds between the support beams and purlins had cracked and the rear of the trailer actually slid off the wheels while traversing the curve, a very different "load shift"! Metallurgical testing revealed that the cracks occurred in untempered martensite that formed in almost all the welds and adjacent heat-affected zones.

Case Study 10.5

Weld Corrosion

In Case Study 11.4, we will see corrosion failure of welds in a large steam generator that produced electricity by burning trash. The anode in the corrosion cell was Widmanstatten ferrite, a microstructure you didn't learn about when studying heat treating. It is formed only in carbon steels containing about 0.30% C and only for specific cooling rates through a critical temperature range. Unfortunately, SAE 1030 steel was the material used to manufacture the preheating chamber for steam generation. The design consisted of pipes (headers and footers) welded to very long vertical tubes. Combustion gases passed by and preheated the tubes, which contained flowing water.

When the original economizer had been replaced, a nipple was used between the headers and tubes. The material was from a different vendor, although purchased to the same specifications. This design required SMAW welds between the nipples and header and between the nipples and tubes, each of which had different metal thickness. In the course of welding, the rate of cooling as a result of the welding practice and thickness variations formed Widmanstatten ferrite.

The welded assemblies were put together and placed in service. However, the Widmanstatten ferrite was anodic to the surrounding pipe, nipple, and tubing microstructures, and corrosion caused leakage and failure within a few months. Proper preheating would have compensated for the thickness variations and eliminated the cooling rates that formed the Widmanstatten ferrite, thereby eliminating the corrosion failure.

Case Study 10.6

Casey at Bat

Hitters in most amateur softball leagues use aluminum bats. As in most activities of this type, the league provides the equipment and all the bats are stored in the bat bag between practices and games. When a new bat broke about 12 in. from the grip end, the team manager put it into the bag and later asked a local welding shop if it could be repaired. It was arc welded, ground smooth, and new tape was applied to the grip area, hiding the weld. The bat held up for a few hits. Then a batter was warming up when the end of the bat came loose, flying into the stands where it struck and injured a fan. Visual inspection showed that the repair should not have been made because of the internal porosity of the weld fusion zone, a common problem in aluminum welding.

Case Study 10.7

Chair Failure

Drafting chairs are frequently used for light assembly tasks. These stools swivel but do not tilt, have polypropylene seats, and are advertised as heavy-duty, strong stools. The design for the chair was adequate, but failures occurred after short-term use, sometimes leading to injury of workers using the chairs. Examination of the broken welds showed that arc welding of thick and thin sections weakened the thin members and that failures of these weakened areas occurred by cracking and shear.

10.3.7   Weld Testing

Testing of welds is necessary to ensure that the joint strength equals the design strength. However, destructive testing of welds is valid only when determining how best to make a weld. For example, a welding company was hired to replace sections of a water tower that had corroded. They made SMAW test welds using several different electrodes, then sectioned them, measuring hardness and examining microstructure to match the base metal as closely as possible, thereby minimizing future corrosion potential. Destructive testing methods that are common include metallography, hardness, tensile and bend testing, impact testing, and fracture toughness, which we were introduced to in Chapter 1. Destructive testing is the exception, however, rather than the rule, because we want to have the assurance that the actual weld will meet our requirements. Therefore we are most interested in nondestructive means to evaluate how good our welds are.

The most common nondestructive test is a **visual inspection**. Such an exam would not have revealed the incomplete penetration in Case Study 10.3, but would have prevented the problems exhibited in Case Study 10.7. There are a number of **nondestructive tests** (frequently referred to as **NDT**) that we can use to determine how good a weld is. Surface cracks can be found by visual exam, liquid dye penetrants, magnetic particle testing, or eddy current testing. Internal defects such as porosity can be found by radiography or ultrasonic testing. Each of these methods has advantages and limitations. Following are more detailed descriptions of these NDT tests.

Dye penetrant test. A liquid is applied to the surface, allowed to penetrate, and excess liquid is wiped away. A developer brings remaining penetrant that has seeped into surface defects to the surface where it is visible under ultraviolet light.

Magnetic particle test. When a ferromagnetic metal is magnetized, any discontinuities create a north-south pole region that strongly attracts magnetic particles. Although the method detects very fine surface flaws, it is limited to ferromagnetic metals.

Eddy current test. Eddy currents are surface currents induced when a metal is brought into an electric field. Surface discontinuities alter the eddy currents, which can be measured with a search probe. Microstructural and normal surface variations limit this technique.

Radiography. When X-rays are passed through a metal, density changes alter the image exposed on film in much the same way as dental X-rays detect cavities. Porosity is readily diagnosed with this method, but stringent safety precautions must be followed.

Ultrasonic test. Internal defects reflect sound waves that are passed through a metal. These waves are received and converted to electrical signals that can be displayed on an oscilloscope. Although defects are easily identified, their size and type are not easy to interpret.

Summary

Our interests in the behavior of metals in their final application requires understanding the methods that we use in joining metals together. Mechanical joining, adhesive bonding, and metal joining were covered in this chapter. Joint design, adherend composition, filler composition, cleanliness, fixturing, and method of application are important considerations common to all the joining methods. Mechanical joining traditionally includes riveted and bolted joints. Adhesive bonding has many advantages, such as stress distribution over large areas, but presents other problems, such as limited service temperatures. Adhesive materials are polymers that are classified as chemically reactive, evaporative, or hot melt types. In metal joining, which includes soldering, brazing, and welding, alloying actually takes place between the metals being bonded and the filler metal. Welding methods include chemical and electrical. Metallurgy in welding is important because the weld metal has a cast microstructure and the adjacent heat-affected zone can have significant property changes. Nondestructive test methods are needed to evaluate the effectiveness of joining. These include visual inspection, dye penetrant testing, radiography, ultrasonic testing, and others.

Terms to Remember

adherend

adhesive

arc welding

batch process

bearing stress

bolt

brazing

butt joint

chemically reactive

cleanliness

clearance

continuous process

dewetting

dip brazing

dye penetrant test

eddy current test

electrode

electron beam welding

evaporative adhesive

filler

flux

fusion

gas welding

GMAW

GTAW

heat affected zone

hot melt

lap joint

laser welding

magnetic particle test

manual processes

mechanical joining

NDT (nondestructive test)

oxyacetylene weld

pitch

plasma welding

preforms

projection weld

radiography

resistance welding

rivet

seam weld

SMAW

soldering

stickscrew

Thermit weld

thermocompression weld

ultrasonic test

visual inspection

wave soldering

welding

Problems

1. In your own words, explain the different types of stresses on a riveted lap joint.
2. What is the shear stress on a single 0.188-in. diameter rivet in a lap joint subjected to 100 lb tensile load?
3. Why should we design a mechanical joint so that failure will occur in the rivet or bolt?
4. Using Table 10.2, compare the advantages and disadvantages of the types of adhesives listed.
5. In your own words, explain how microalloying affects the solderability of an adherend metal.
6. Explain the role of cleanliness in soldering, brazing, and welding. How can we control the joint cleanliness?
7. Explain how automatic sprinklers and thermal alarms utilize metal-joining technology.
8. Compare the methods for soldering or brazing from the viewpoint of productivity.
9. Should we design welded joints so that failure occurs in the joint? Compare your answer to that for Question 3.
10. What are some of the safety precautions that should be considered in oxy-acetylene and arc welding?
11. In your own words, compare the microstructure of the weld zone, the heat-affected zone, and the adherend metal for arc-welded SAE 1020, 1040, and 1060 steel.
12. How would electrode selection affect your answer to Question 11?
13. Compare the advantages and disadvantages of three new technology welding processes.

11

Corrosion and Corrosion Protection

Corrosion is responsible for substantial annual economic loss and we must be aware of how it occurs and how we can prevent it both in our own manufacturing as well as for the products we produce. **Corrosion** is simply the destruction of metal by chemical or electrochemical reaction with the environment. It can lead to gradual failure because of weakening through geometric loss, such as the deterioration of our highway bridges in the United States. But it can also cause sudden and catastrophic failure, with little corrosion product formed. In this chapter, we will examine what causes corrosion and what we can do to prevent it or slow it to extend the lifetime of metals.

11.1   The Nature of Corrosion

The chemical or electrochemical reaction that corrodes metals forms a corrosion product, usually an oxide. Damage caused by the corrosion, however, also depends on the rate of corrosion. Usually, the formation of the corrosion product slows the rate because the corrosion product isolates the metal from the environment. If the corrosion product is adherent, then the overall corrosion rate is low and we think of the metal as corrosion resistant. The best example of this is aluminum, which forms a tightly adherent oxide, Al_2O_3, that is not only adherent but insoluble in water and many chemicals. The worst example of corrosion resistance, of course, is rusting of steel and other iron-base alloys. Rust, the technical name for the corrosion product of iron, is not adherent, so fresh, underlying surfaces are continually exposed to additional corrosion.

Rust is not adherent because it forms in a two-stage reaction, first forming ferrous hydroxide, $Fe(OH)_2$, which is soluble in water. Hydration away from the solid surface then precipitates ferric oxide, Fe_2O_3, which is the flaky, orange material that we associate with rust. It follows that the corrosion rate for rusting is high.

The mechanism for corrosion is **electrochemical**, involving an anode and a cathode in every corrosion situation. Referring to Figure 11.1, it is the anode that corrodes, with the anodic reaction for a metal M:

$$M \rightarrow M^{++} + 2e^-$$

There are two possible cathodic reactions. General attack is associated with

$$\frac{1}{2}O_2 + H_2O + 2e^- \rightarrow 2OH^-$$

A pitting type attack is associated with the reaction

$$2H^+ + 2e^- \rightarrow H_2$$

We can conclude from these electrode relations that both the metal and the environment affect the corrosion resistance in any particular instance.

In order for corrosion to take place, we must have a **galvanic cell**, which requires a cathode and anode in electrical contact with each other, usually through moisture. There are three types of galvanic cells that we encounter: composition, stress, and concentration cells (shown in Figure 11.2).

It is always the **anode** that corrodes because the surface metal atoms ionize, releasing electrons that travel to the **cathode**, producing a corrosion current. In a **composition cell**, which metal corrodes will depend on individual ionization characteristics. For pure metals, we can measure the voltage generated by ionization in a

Figure 11.1
Electrochemical nature of corrosion

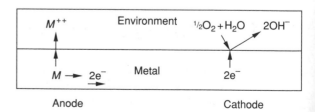

Figure 11.2
Types of galvanic cells:
(a) composition cell,
(b) stress cell, (c) oxygen
concentration cell

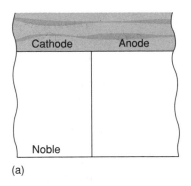

(a)

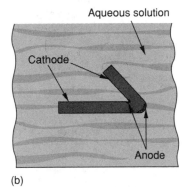

(b)

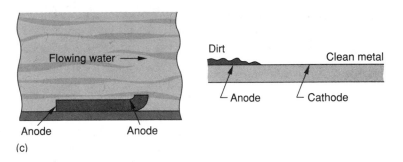

(c)

standard cell. The voltage or electrode potential can be listed numerically, forming an electromotive force (EMF) series that indicates which metal will be anodic in a composition cell. Metals that do not corrode, such as gold and platinum, are referred to as **noble** metals. The EMF series is limited, however, in part because it cannot include alloys. Generally, it is more useful to construct a galvanic series, such as that for flowing seawater, given in Table 11.1. A **galvanic series** applies only to the specific environment, but more accurately reflects what is anodic when two metals are coupled in that environment.

Table 11.1
Galvanic series in seawater

Anodic
Magnesium and its alloys
Zinc
Beryllium
Aluminum alloys
Cadmium
Low-alloy steel
Cast iron
Tin
Brass
Copper
Bronze
Stainless steel types 410, 416 (active)
Lead-tin solder
Lead
Cupronickels
Silver braze alloys
Monels
Stainless steel types 304, 316 (passive)
Titanium
Cathodic

Case Study 11.1

Composition Cell Corrosion

Whenever there is an industrial fire or flood, we must assess the damage in order to determine the extent of loss for insurance coverage. Corrosion is a common cause of such losses. Assessment of damage has been difficult for electronics, sometimes because of rapid obsolescence of equipment, but also because of lack of information on how damaging the corrosion product is. One perplexing problem that has been repeatedly encountered is a white deposit on printed circuit boards, such as that shown in Figure 11.3. For a long time, the nature of this deposit was unknown, leading to full-value damage claims. The deposit was conductive, but could be cleaned off with a combination of mechanical scrubbing and solvent cleaning. After one rather large loss, where the deposit could be extracted and identified by X-ray diffraction analysis, it was found that the deposit was almost pure SnO_2, a corrosion product of the microcomposition cell of the Pb-Sn solder eutectic microstructure. The lead-rich phase is cathodic to the tin-rich phase (in agreement with Table 11.1, as we would expect).

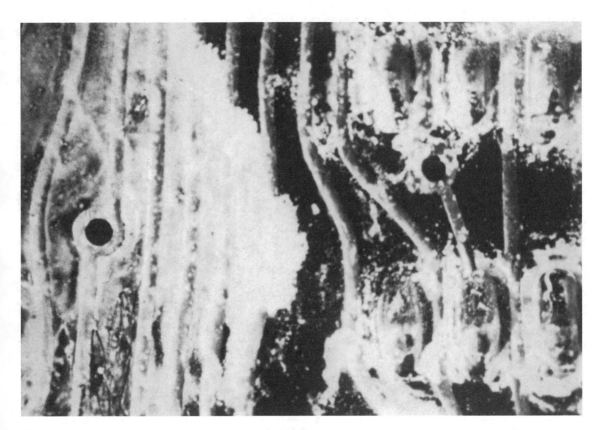

Figure 11.3
White corrosion deposit on a printed circuit board

Case Study 11.2

Stress Cell Corrosion

Modern surgical techniques make use of many specialty metals, particularly corrosion-resistant metals. Arterial catheters, for example, are made of **austenitic stainless steel** wire wound on a mandrel to form a coil. There are two safety wires located inside the coil, which is only $\frac{1}{32}$ in. in diameter. In Chapter 9, we learned about deformation, but here we only need to recognize that deformation stresses the metal. Proof testing of the catheters in a saline solution caused failure of the outside coil, shown in Figure 11.4. This fracture is an example of stress corrosion failure, with the stress being a combination of residual and applied stresses of the proof test and the environment being the saline solution. Incorporation of a stress relief anneal after winding the coils eliminated the problem.

Figure 11.4
SEM micrograph of an arterial catheter that failed in proof testing (150×)

Case Study 11.3

Oxygen Concentration Cell Corrosion

There are many miles of buried pipelines in the world that carry flammable gases, oil, water, and many other fluids. There is a frequent misconception that stainless steel pipe will last much longer than the much more economical mild steel pipe. This was the mistake made when a local contractor installed an underground drainage pipe above a concrete room that housed electrical equipment for a hospital complex. The contractor followed correct procedures for burying pipe, laying coarse stone beneath the pipe and fine stone and sand above the pipe. In only one year, however, a leak was detected and the pipe was dug up, revealing a pit about $\frac{1}{16}$ in. in diameter that penetrated through the top of the pipe. The normal causes for pitting of stainless steels, such as environmental halides, were not found. The process of elimination left only an **oxygen concentration cell** as an explanation for the corrosion. In this case, oxygen was more plentiful at the bottom of the pipe where it was surrounded by coarse stone, making it cathodic. The finer stone and sand at the top of the pipe made it anodic, leading to the pitting corrosion.

11.2   Corrosion Rate

Although we must know why corrosion occurs, the bulk of corrosion problems really are caused by the rate at which corrosion takes place. Some metals, like aluminum, are anodic but form an adherent film that protects against further corrosion. The corrosion rate, therefore, is very low and we think of aluminum as corrosion resistant. In many respects, the relationship between corrosion and corrosion rate is analogous to equilibrium and whether or not equilibrium can be achieved in the real world.

In order to understand how fast corrosion will occur, we must remember the electrochemical nature of corrosion. The corrosion cell has an anode and a cathode; these electrodes are not at equilibrium when an electrical current flows to or from its surface. When current flows away from the anode, the anode becomes more cathodic and when the electrons flow to the cathode, the cathode becomes more anodic. We describe the potential change (volts) caused by the current flow by the term **polarization** and describe the type of control by which electrode polarizes more. Figure 11.5 illustrates the polarization process schematically and the types of control we normally see. An example of mixed control is a zinc anode and copper cathode.

The corrosion potential is the voltage at which polarization is complete and the corresponding (maximum) current is called the corrosion current. The corrosion rate is determined by the **current density**, that is, the corrosion current divided by the anode area. Current density is a particularly important consideration when there is a large value of the ratio of cathode area to anode area. As we can see in Figure 11.6, current density (J) and, thus, the corrosion rate are very high in such circumstances and pitting type corrosion occurs (such was the case of the buried stainless steel pipe in Case Study 11.3).

11.3   Passivity

Closely related to polarization and corrosion rate is **passivity**, the passive behavior of metals. A passive metal is one that is active in the EMF series, yet corrodes at a very low rate. Such a metal polarizes according to Figure 11.7. Some metals are nat-

Figure 11.5
Polarization: (a) mixed control,
(b) cathodic control, (c) anodic
control

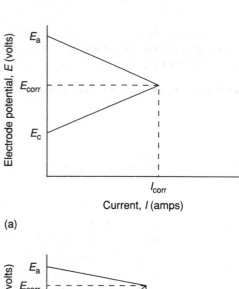

(a)

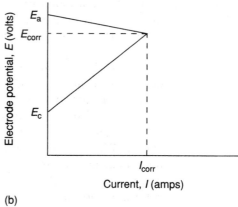

(b)

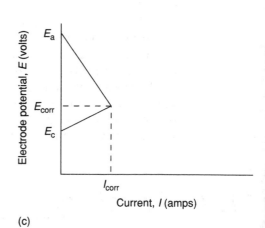

(c)

Figure 11.6
Effect of the ratio of anode area to cathode area on corrosion current density: (a) equal anode areas, (b) large anode area and small cathode area, (c) large cathode area and small anode area

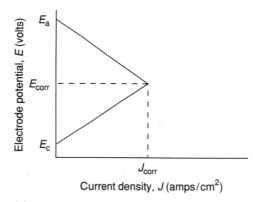

(a)

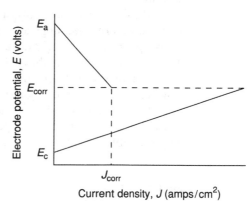

(b)

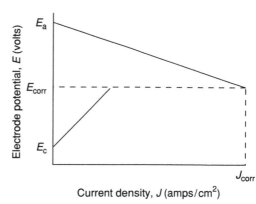

(c)

urally passive, such as chromium, but other metals, particularly steel, which is so important commercially, can be passivated in certain environments. Passivating substances for iron include chromates, nitrites, and molybdates. We usually attribute the **passivation**, or improvement of corrosion resistance, to development of a thin protective film that is impermeable to the corrosive environment.

The natural passivity of chromium can be imparted to alloys that contain it, enabling development of a series of alloys we call **stainless steels**. Figure 11.8 shows the dramatic reduction in corrosion rate of Cr-Fe alloys with increasing chromium concentration. All stainless steels contain at least 12% chromium by weight. We also find that the corrosion resistance is related to crystal structure and, in certain instances, to the crystal face that is exposed. For example, austenitic stainless steels (containing Ni) have better corrosion resistance than ferritic stainless steels, and the orientation of silicon wafers in microelectronic processing determines the chemical etching rates.

Passivators are used in water treatment for heating and air conditioning maintenance in large buildings and are present in the well-known rust-inhibiting paints. These passivators are chemically reduced during polarization over large cathodic areas to an extent that they form the passive film at small anodic areas. When the film is complete, the anode acts as a cathode, preventing further polarization and slowing any further reduction of the passivator.

One note of caution — halogen ions, particularly chlorides because of their extensive use commercially, penetrate and destroy the passive film easily. This action establishes an active-passive cell with small anode and large cathode, thereby

Figure 11.7
Passivation polarization

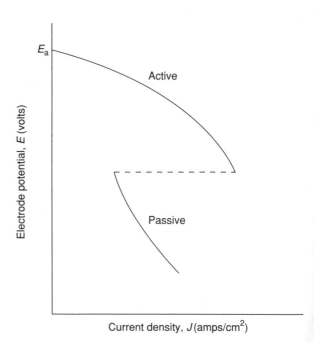

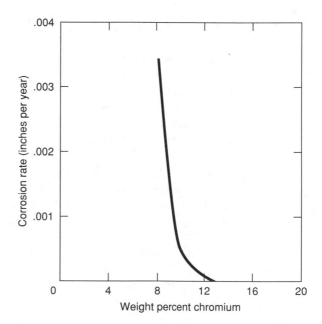

Figure 11.8
Influence of chromium content on corrosion rate of Fe-Cr alloys

producing high current density at the anode and rapid, pitting type corrosion. We always look for sources of halides whenever unexpected pitting corrosion causes failure in service.

11.4 Corrosion Protection

After examining the factors that influence corrosion and corrosion rate, we have the foundation to begin a corrosion protection program. In addition, economics play a large role in our decision, for it may be easier to replace certain equipment than to make it out of more expensive but corrosion-resistant materials. The common methods that we use for corrosion prevention and protection are

material selection
material design
inspection
coatings
cathodic protection
environmental control

Let's look at these in more detail in light of what we have learned about corrosion.

11.4.1   Material Selection

Material selection is our first and foremost consideration so that we choose the right material for our application. There are many applications where we have little choice other than the expensive one. For example, chemical machining utilizes strong chemicals that attack the workpiece, but the framework of the machine must endure these same chemicals; the only material for this purpose is titanium. Likewise, marine applications need to withstand the deleterious effects of the chloride environment, so we must use an alloy such as the monels (see Table 11.1). In many instances, however, we can substitute plastic parts that will not deteriorate in the environment. Most importantly, we have to think about corrosion before making decisions on material selection. In Case Study 11.3, there was no real understanding of corrosion protection, demonstrated by the choice of stainless steel for the buried pipe.

Closely related to material selection is the choice of material processing, which must also consider corrosion effects. Perhaps the most common example of the need to select the right processing is the **sensitization** of stainless steels. When heat-treated in the range of 800–1500°F, stainless steels with carbon contents greater than about 0.03% will have chromium carbide precipitate at grain boundaries (shown schematically in Figure 11.9), depleting the chromium content below the 12% needed for corrosion resistance. The result is intergranular corrosion. Stainless steels are stabilized by adding carbide formers or by producing very low carbon content stainless steel.

Case Study 11.4

Corrosion of a Preheater for Steam Generation

Conservation and recycling are popular concepts, but the technology, costs, and problems in establishing programs are ongoing, expensive, and complicated. If we consider garbage as an example, landfills that have been the norm are no longer popular because of hygienic problems and rising costs. One alternative to landfills is incineration, which generates electricity as a by-product of the steam produced. Two basic incineration processes have evolved, the first burning all garbage with a low heat content fuel, and the alternative using a pretreatment separation process that enriches the heat content of the fuel (see Case Study 6.2). Figure 11.10 illustrates the typical flow process when there is no pretreatment. In this process, available heat does not satisfy the requirements of normal boilers for steam generation, so a lower circulation rate is combined with preheating in three sections: the economizer, steam generator, and superheater. These 60-ft high sections are made of headers and footers connected through vertical pipes, with water flow counter to gas flow. Obviously, high-temperature corrosion is an important consideration that is monitored constantly. The original installation performed satisfactorily for 8 years, when progressive corrosion of the economizer made it more economical to replace it, even at the cost of nearly $1 million.

Figure 11.9
Sensitized austenitic stainless
steel: (a) microstructure of
austenitic stainless steel,
(b) composition of Cr across
section A–A

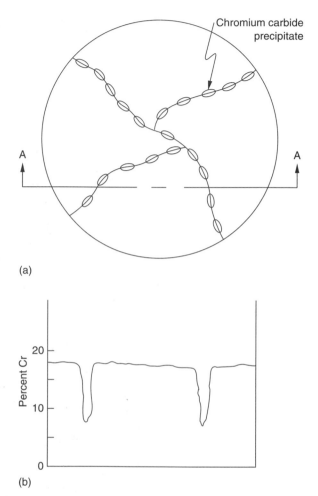

The new economizer was fabricated somewhat differently than the original, with nipples required between the headers and footers and the vertical pipes. Such configuration increased the number of welds, the variable composition of each part, and the thickness differences at each weld, making the welding process more difficult and more critical (see Chapter 10). Significant corrosion occurred within a few months, causing leaks at the nipple–vertical pipe weld locations. Metallurgical failure analysis showed that such corrosion was exacerbated by two weld-related problems. First, excess weld metal projecting into the piping altered the flow pattern and created an oxygen concentration cell and, second, the cooling rate of the weld metal caused by different thicknesses of the metal being joined and the particular alloy content (SAE 1030 steel). (The welding practice led to an unusual microstructure known as Widmanstatten ferrite, which is anodic to the more common ferrite and cementite surrounding the weld.) Replacement of the relatively new economizer was required.

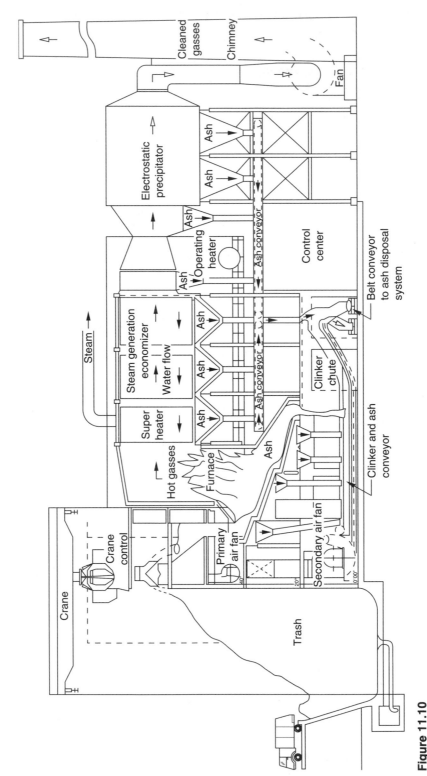

Figure 11.10
Typical flow process for steam generation from trash incineration

11.4.2 Material Design

There are two main issues we face in **material design** for corrosion protection. First, we want to avoid stresses, both residual and applied, when operating in any corrosive environment. An example of this is the stress relief anneal given the stainless steel coils in Case Study 11.2. Applied stress is more difficult to deal with, unless we consider it in the design phase, reducing the applied stress by altering the cross section. We also can avoid dissimilar metal contact, crevices, and designs that lead to oxygen concentration cells wherever possible. One manufacturer of pressure transducers found that the surface finish of their extra low carbon austenitic stainless steel, produced by grinding, caused it to corrode in a kidney dialysis machine. The problem was attributed to a contaminated grinding wheel.

Case Study 11.5

Design of a Hot-Water Combustion Chamber

Heating and air conditioning in many public buildings depend on boilers. In most instances, the water that is recirculated is treated daily with chemicals to prevent corrosion. Two identical boilers that supplied heat and air conditioning to a local police headquarters and the adjacent civic auditorium developed leaks caused by corrosion within 5 years of installation, despite continuous satisfactory daily water tests. Failure analysis showed that the corrosion was related to oxygen concentration cells established by the design and manufacture of the walls to the combustion chamber.

Most boilers have heating tubes that pass the water through the combustion chamber, similar to the steam generation in Case Study 11.4. In the design of these boilers, however, insulated walls of the combustion chamber were replaced with "water walls" — two steel surfaces containing water that would serve as insulation and provide preheating before passing through the tubes. Corrosion such as that shown in Figure 11.11 caused leaks at all four lower corners. The design details were reviewed, and it became apparent that ease of fabrication processing was the determining factor and no consideration was given to the relation of the design to corrosion prevention. Figure 11.12 shows the inner wall and base construction of these leaky lower corners. The intersection of three surfaces in this area plus the overlap caused by the bending of the inner wall to form the bottom created turbulent flow, with pockets of stagnant water, leading to oxygen depletion and erosion corrosion by the concentration cell. After repeated costly repair attempts, the boilers were eventually replaced with two new boilers of more conventional design.

Figure 11.11
Corrosion at a bottom corner
of the boiler

11.4.3   Inspection

A major problem with corrosive failure is the damage that can ensue. **Inspection** for corrosion, cleaning, and/or replacement before failure occurs can be an effective means of corrosion protection. For example, the National Fire Protection Agency calls for semiannual corrosion inspection of automatic sprinkler heads in their specification NFPA 13, *Care and Maintenance of Sprinkler Systems*. Many cases of flood damage that occurred because of corrosive failure of sprinklers were needless and could have been prevented through inspection.

SHOP NOTE: CORNER TO BE STRESS RELIEVED
AFTER WELDING

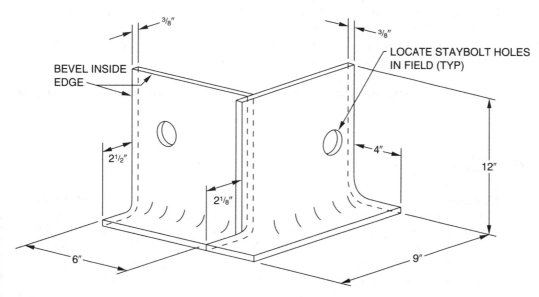

Figure 11.12
Boiler wall corner design, combustion side

What do we look for in an inspection? Well, in the case of stress corrosion, there is little we can do, but for other types of corrosion, all we need to do is look for evidence of corrosion product. This could be the rust formed on steel, the white deposit in Case Study 11.1, or the bluish patina formed on the brass of automatic sprinklers.

Case Study 11.6

What Inspection Could Have Prevented

Reconstruction of an industrial accident, which resulted in an injury, revealed a corrosion problem that would have been prevented if detected by inspection. While rinsing parts that had been electroplated, a worker was scalded by the rinse water. Rinse water temperature was controlled by a thermostat that had a stainless steel tube brazed to a stainless steel bellows. When temperature decreased, the silicone oil inside the tube would contract, causing the bellows to contract and turn on the heat. Conversely, higher temperatures would expand the oil and extend the bellows, causing the heat to shut off. The scalding accident occurred because the braze, an alloy of silver, copper, and zinc, corroded and caused a pit because of the large cathode to anode area. The silicone oil leaked out and higher heat never caused the bellows to expand.

During the accident investigation, other thermocouples of the same design in use in the department were observed to have obvious corrosion product formed in the braze area. There is no doubt that a proper inspection and replacement program could have been in place that would have prevented the worker's injury.

11.4.4   Coatings

Coatings provide corrosion protection in a number of ways. Separation of the metal from the corrosive environment by a coating is the most common and can be done with metal, ceramic, or polymer coatings. In all cases, coatings must be tightly adherent or moisture can penetrate and establish galvanic cells, resulting in substantial damage beneath the coating before it is apparent by external inspection. Of course, coating adherence is promoted by clean metal surfaces and application of a primer coat before application of the protective coating. Primers often contain corrosion **inhibitors** such as zinc chromate that act as passivators as well as barriers to the environment.

By and large, the most common coatings are organic, including paints, varnishes, lacquers, and polymeric coatings of many types. These coating surfaces must be completely free of porosity or enhanced pitting type corrosion can occur. Therefore, multiple applications are recommended and thickness is important to prevent premature breakdown as the paint deteriorates or weathers. Selection of the protective coating is important and can be influenced by the expected environmental exposure, temperature, and service use. Most coatings are thermosetting polymers; for example, we frequently use polyurethane and epoxies for floors, phenolics and polyesters for tank linings, and silicones for high-temperature exposures.

Coatings can serve another role as well. Metal coatings can be either cathodic or anodic to the metal being protected. When coating with noble metals, which are cathodic to the metal being protected, the surfaces must be completely free of porosity or enhanced pitting type corrosion can occur. Common examples of noble metal plating are gold-plated fingers on printed circuit boards for microelectronics and terneplate, a Pb-Sn alloy coating for steel gasoline tank applications. If an anodic metal is used for coating, the situation is very different if a porous surface occurs and the underlying metal is exposed to a corrosive environment. The anodic coating will corrode, protecting the underlying metal. **Galvanizing**, or coating with zinc, is the best known example of this type of coating, but the use of cadmium-plated nuts and bolts is also commonplace in manufacturing. The difference in the behavior of these two types of metal coatings is shown schematically in Figure 11.13.

Figure 11.13
Influence of porosity in protective metal coatings: (a) accelerated attack, (b) cathodic protection by sacrificial coating

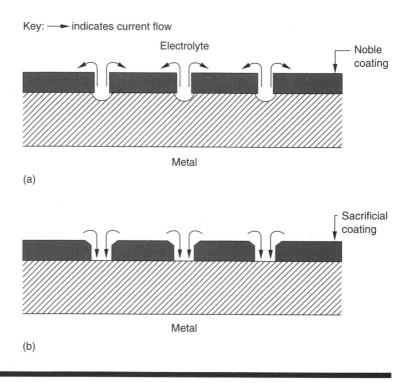

Key: ➝ indicates current flow

Electrolyte

Noble coating

Metal

(a)

Sacrificial coating

Metal

(b)

Case Study 11.7

A Corrosion-Resistant Garden Sprayer

What could be more corrosive than a pesticide used for garden insect control? A garden sprayer that is corrosion resistant therefore has sales appeal. However, to use such appeal to advantage, we really need to understand what corrosion resistance is. One manufacturer of sprayers learned this lesson the hard way, having to recall the sprayers after many of them exploded. Failure analysis showed that the seam welding of galvanized steel (the corrosion resistance) evaporated the zinc, of course removing the corrosion protection. Where the welds were in critical spots, such as the intersection of the tubular section with the bottom, lack of corrosion resistance combined with oxygen concentration cells and the extreme corrosiveness of the contents to cause explosive failure when pressurized by the hand pump.

11.4.5 Cathodic Protection

Cathodic protection against corrosion is achieved by supplying electrons to the anodic metal to be protected, polarizing it to the same potential as the cathode. By suppressing any current flow from the anodic metal, corrosion is prevented. There are two ways to cathodically protect a metal from corrosion. We have already seen the first in galvanized steel coatings, because galvanic coupling does provide electron flow from the zinc to the steel, protecting the steel sacrificially. Coatings are unnecessary, however, for we can use **sacrificial anodes** that are electrically connected to the metal to be protected. This is common practice to protect the steel hulls of ships and, as shown in Figure 11.14, to prevent internal corrosion of domestic water heaters.

Current can also be impressed by an external power supply, connecting the negative terminal to the metal to be protected and the positive terminal to an anode such as graphite. An example of this type of protection, shown in Figure 11.15 for steel pilings of ocean piers, uses large rectifiers and graphite anodes. Maintenance and inspection of these systems is, of course, mandatory.

11.4.6 Alteration of Corrosive Environment

The last method of corrosion protection we will look at is environmental control. There are many ways that we can change the environment, such as lowering temperature wherever possible (any chemical reaction proceeds at higher rates as temperature increases), decreasing water flow to avoid erosion corrosion (see Case Study 11.5), removing oxygen by deaeration or addition of oxygen-scavenging inhibitors, altering concentration such as pH control or chloride reduction in boiler water treatments, and by passivation with inhibitors. A special case of inhibitors is the adsorption type, for example, organic amines. These compounds adsorb onto the surface of the metal to be protected, providing a barrier to corrosion.

Figure 11.14
Cathodic protection by
sacrificial anode in a
domestic hot water tank

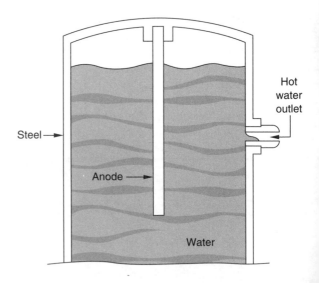

Steel→

Anode →

Hot
water
outlet

Water

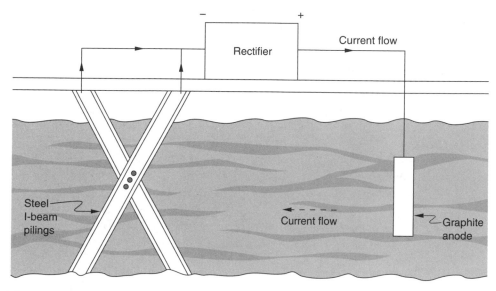

Figure 11.15
Cathodic protection of steel pier pilings using impressed currents

Summary

Corrosion, the electrochemical destruction of metal, is caused by galvanic cells in which the anode corrodes. There are three types of galvanic cells: composition, stress, and concentration. The electrical potential depends on metal composition, described qualitatively by the galvanic series, with noble metals acting as cathodes. In **stress cells**, either residual or applied stressed areas are anodic. The most common **concentration cell**, in which the electrolyte is concentrated in one area, is the oxygen concentration cell where the area depleted in oxygen is anodic. When current flows, the electrodes are polarized until the electrical potential is the same for anode and cathode; current at this value is known as the corrosion current. Corrosion rate is determined by the corrosion current, with direct correlation between corrosion rate and anode current density. Polarization is interrupted in certain conditions, and active metals corrode at lower rates. This phenomenon is known as passivity and can occur naturally, as for chromium and stainless steels, or can be affected by passivators such as zinc chromate in the environment. Corrosion prevention and protection is best practiced by material selection, material design, inspection, protective coatings, cathodic protection, and environmental alteration.

Terms to Remember

anode

austenitic stainless steel

cathode

cathodic protection

coatings

composition cell

concentration cell

corrosion

current density

electrochemical

galvanic cell

galvanic series

galvanizing

inhibitors

inspection

material design

material selection

noble

oxygen concentration cell

passivation

passivity

pH

polarization

sacrificial anode

sensitization

stainless steel

stress cell

Problems

1. Define in your own words what corrosion is and what factors affect corrosion of metals.
2. Identify the anode in each of the following galvanic cells and explain your answers.
   a. a two-phase microstructure of a lead-tin eutectic alloy solder
   b. a silver-copper-zinc braze joining SAE 1040 steel
   c. a silver-copper-zinc braze joining stainless steel
   d. an aluminum rivet joining aluminum of the same composition
   e. a buried pipe with air space above caused by settling
3. Explain why stress corrosion can lead to sudden cracking and catastrophic failure.
4. Can heat treatment affect the corrosion resistance of high-strength low-alloy steels? Explain your answer.
5. Explain in your own words what passivity is and how we can use it to advantage.
6. Explain why failure analysis of pitting type corrosion always begins with chloride analysis.
7. Explain what intergranular corrosion of stainless steel is. How might you prevent it?

8. Explain what factors you must consider in selecting metals for applications where corrosion might be a problem.
9. Construct a step-by-step process description for painting exposed steel columns in your plant.
10. Explain why sheet metal screws for electronic cabinets are cadmium-plated.
11. Explain how stress corrosion can be reduced or eliminated.
12. Describe three methods that alter the environment to control corrosion.
13. Outline a corrosion inspection program for an electronics manufacturing plant.

Glossary

Activation energy   Energy barrier that must be overcome in order for an action or reaction to occur

Adherend   The surface to which an adhesive bonds

Adhesive   A substance that bonds two surfaces together

Age hardening   *See* precipitation hardening

AISI/SAE   Acronym for American Iron and Steel Institute and Society of Automotive Engineers who specify standards

Alloy   A metal containing more than one element

Alnico   Alloys containing aluminum, nickel, and cobalt and used for permanent magnets

Anisotropy   The quality of having properties that depend on the direction in which they are measured

Anneal   *See* full anneal, process anneal, and stress relief anneal

Anode   The electrode in a galvanic cell that oxidizes, giving up electrons

API   Acronym for American Petroleum Institute

Arc welding   A welding process that uses the energy of an electric arc for heating

ASTM   Acronym for American Society for Testing and Materials

Atom   The smallest particle of an element that combines with similar or dissimilar atoms to form matter

Atomic bonding   The attraction that combines atoms to form matter

Atomic number   The number corresponding to the total number of electrons or protons that occur in an atom

Atomic packing factor   The fraction of a unit cell occupied by atoms

Atomic weight   The weight of one mol (6.023×10^{23} atoms) of an element that depends on number of protons, electrons, and neutrons present in the atom

Ausforming   Thermomechanical deformation of austenite before transformation to refine penultimate grain size

Austempering   A heat treatment of steel whereby austenite is quenched to below the knee of the TTT diagram and held isothermally to form bainite

Austenite   The fcc high-temperature phase of iron or steel

Austenitic stainless steel   Stainless steel that is fcc because it contains nickel

Austenitization   Heating iron or steel above the eutectoid temperature to form austenite

Babbitt metal   Antifriction bearing alloys based on lead or tin

Bainite   Very fine, feathery microstructure of ferrite and cementite formed isothermally

Basic oxygen furnace   A steel-refining furnace that uses oxygen in processing

Batch process   Any process that is constrained to a fixed amount

Bearing stress   The compressive stress on metals that are mechanically joined

Beryllium-copper   A precipitation hardenable copper alloy that contains beryllium

Billet   A semifinished metal section deformed by rolling, forging, or extrusion

Binary alloy   An alloy of two elements

Blanking   Cutting a metal sheet with a die to form metal parts

Blast furnace   A furnace in which a semicontinuous process for converting iron ore into pig iron takes place

Body-centered cubic (bcc)   Cubic arrangement of atoms in each corner and in the center of the unit cell

Bolt   A threaded fastener used with a nut to mechanically join two or more parts

Bohr magneton   A unit of magnetization

Braid   A woven covering for electrical cable

Brass   The alloy of copper and zinc

Brazing   A metal-joining procedure involving a dissimilar metal braze alloy that fuses at temperatures above about 800°F

Bronze   The alloy of copper and tin

Bull block   A wire-drawing machine that employs a rotating capstan to pull wire through a die

Butt joint   A joint made by joining parts end to end

Cable   A flexible wire rope made by twisting strands of wire together in a geometric pattern

Capstan   The tapered drum that rotates, providing force to draw wire through a die

Carbides   Metallic compounds formed with carbon that have high melting points and high hot hardness

Carburizing   A heat treatment or atmosphere that promotes carbon formation and diffusion into a metal at high temperatures

Cartridge brass   An alloy containing 70% copper and 30% zinc

Casting   (1) The shape made by pouring molten metal into a mold and letting it solidify; (2) the foundry process of making the shape

Cast iron   An alloy of iron, carbon, and silicon that contains more than 2% carbon

Cathode   The electrode of a galvanic cell that accepts electrons and does not corrode

Cathodic protection   A means of impressing a current on an electrode to make it cathodic

Cementite   The name for iron carbide, Fe_3C

Charpy impact test   A kinetic energy impact using standard test and standard test bar

Chemically reactive   A type of adhesive that requires chemical reaction to join materials

Chill zone   Fine-grained metal microstructure of first metal to solidify from the liquid

Cleanliness   Freedom from contaminants that might interfere with processing

Clearance   The space between component parts

Closed die forging   A deformation process that shapes a metal by compression between shaped dies

Coatings   Thin layers of protective material

Coercive force   The magnetic field required to demagnetize a ferromagnetic material from remanence

Cold working   Deformation carried out at temperatures below the recrystallization temperature of a metal

Cold work tool steel   Tool steel that is used for cold-working tools

Columnar grains   Column-shaped grains that form by growth of nuclei at the end of the chill zone during solidification

Composition cell   A galvanic cell based upon the electrochemical difference between two electrodes

Compression   The squeezing action caused by forces directed at each other

Concentration cell   A galvanic cell based upon the electrochemical difference caused by concentration of the electrolyte

Conductivity   The ability of a material to carry electrical current; the reciprocal of electrical resistivity

Constantan   A nickel-copper alloy used for thermocouples

Continuous process   A process that continuously produces material

Cooling curves   Heat-treating diagrams that describe the change in temperature with time during cooling from a high temperature

Coordination number   The number of nearest neighbors of an atom in a unit cell

Core loss   Energy losses in magnetization cycles of electrical transformers

Coring   Segregation observable in cast microstructures

Corrosion   The environmental destruction of a metal by oxidation

Covalent bond   The directional bonding of atoms by the sharing of valence electrons

Critical resolved shear stress   The minimum stress required to cause slip

Crucible   The container in which metal is melted

Crystal structure   The long-range geometric array of atoms bonded together in a solid

Cupronickel   A copper-rich, copper-nickel alloy

Curie temperature   The temperature above which ferromagnetism disappears

Current density   The amount of electrical current in a metal as a function of its cross-sectional area

Deep drawing   A forming method that bends and stretches a metal into a desired shape

Defects   Imperfections in the structure of a material

Dewetting   Lack of adherence of solder to the metal being joined

Diamagnetic   Induction of a magnetic field in a material that is lower than the applied magnetic field

Die casting   Casting and solidification of a metal in a permanent mold under pressure

Diffusion   Movement of atoms from a region of higher concentration to a region of lower concentration

Dip brazing   Metal-joining process whereby the fixture of parts being joined is dipped into a bath of the braze alloy

Dislocation   A linear defect in the crystalline lattice

Distribution coefficient   The ratio of the concentration of the solid phase to that of the liquid phase during solidification

Draw bench   Machine for drawing straight rod or wire

Ductile cast iron   Cast iron containing graphite as nodules formed during solidification

Ductile-to-brittle transition temperature   The temperature range where the fracture of bcc metals changes from ductile to brittle as temperature is lowered

Ductility   The ability to deform without fracture, usually reported as percent elongation or percent reduction in area

Dye penetrant test   A nondestructive test for surface weld defects

Eddy current test   A nondestructive test for surface weld defects that uses magnetic particles

Edge dislocation   A linear defect caused by partial insertion of a plane of atoms into a crystal lattice

Elastic deformation   Deformation that is not permanent

Elastic modulus   The ratio of stress to strain during elastic deformation

Electric arc furnace   A melting furnace that employs electrical discharge for heating energy

Electrical steel   An alloy of iron and silicon used for its magnetic characteristics

Electrical contacts   Metal parts that conduct electricity when in mechanical contact

Electrical properties   Material properties such as conductivity or resistivity in the presence of an electric field

Electrochemical   Combination of electrical current flow between an anode and cathode undergoing chemical change

Electrode   The anode or cathode in an electrochemical reaction

Electron beam welding   The welding process whereby the energy is from an electron beam acting in a vacuum

Electrons   Negatively charged subatomic particles that carry electrical current; in an atom, electrons orbit around the positively charged nucleus

Electrostatic attraction   The attractive force of positively charged and negatively charged particles

Electrostatic repulsion   The repelling force of like-charged particles (positive–positive and negative–negative)

Elinvar   An alloy of nickel, iron, and chromium that has a modulus of elasticity that is invariant with temperature

Elongation   The change in length caused by an external tensile stress, used as a measure of ductility

Embrittlement   Brittle behavior caused by an impurity such as hydrogen

Endurance limit   The lowest tensile strength of a metal under fatigue conditions

Equilibrium   The lowest energy or most stable condition

Eutectic   An invariant reaction in which a liquid forms two solids upon cooling

Eutectoid   An invariant reaction in which a solid forms two different solids upon cooling

Evaporative adhesive   An adhesive dissolved in a solvent that evaporates during bonding

Extrusion   A direct compression process whereby material is shaped by squeezing it through a die

Face-centered cubic (fcc)   A cubic crystal lattice that has atoms at the cube corners and in the center of cube faces

Fatigue   Intermittent or cyclic application of stress

Ferrite   A bcc phase of Fe-Fe$_3$C alloys

Ferromagnetic   A material that has a higher induced magnetic field than that which is applied

Ferrous   Containing iron as the main constituent element

Filler   In welding, the metal added during the welding process

Flux   A material that melts and protects metals during heating to the temperature for joining

Forging   A direct compression process where metal is deformed between two dies

Forming   Secondary processes for the final shaping of a metal part

Fracture toughness, K_{1c}.   The critical value of the stress intensity factor necessary to propagate a crack to complete failure by fracture

Fusion   The process of melting

Fuel-fired furnace   A melting furnace that uses combustion of gas or oil for heat

Full anneal   Softening heat treatment where fully recrystallized metal is furnace cooled

Furnace atmospheres   Vacuum or inert gases for protection during heat treatment or gases for addition or reduction during heat treatment

Furnaces   Structures for heat-treating or melting materials

Galvanic cell   An electrochemical cell comprising an anode that reacts and emits electrons and a cathode that collects electrons

Galvanic series   A systematic listing of corrosion potential of metals and alloys in specific corrosion environments

Galvanizing   Coating steel with zinc for corrosion resistance

Gas welding   A welding process that utilizes gas for the source of energy

Gibbs phase rule   A thermodynamic relation between the number of components and phases that can coexist at equilibrium

GMAW   Acronym for gas-metal arc welding

GP-zones   The coherent precipitate nuclei that are responsible for precipitation strengthening of metal alloys

Grain boundary   A surface defect separating single crystal grains of metals

Grain growth   The enlargement of recrystallized grains during continued heat treatment

Grain size   The description of the average size of grains in a microstructure

Graphite flakes   The form of graphite in gray cast iron

Gray cast iron   The iron-carbon alloy that comprises graphite flakes in a steel matrix

GTAW   The acronym for gas-tungsten arc welding

Hall-Petch relation   The equation that describes strength as a function of grain size

Hardenability   The description of the ability of a steel part to be hardened through its thickness by quenching

Hard magnet   A ferromagnetic material that has been magnetized and remains at remanence; also called a permanent magnet

Hardness   The resistance of a metal to indentation

Heat affected zone   Change in microstructure of metal near a welded joint

Heat treating   High-temperature treatment of a metal to change its properties

Hexagonal close-packed (hcp)   A metal crystal system with hexagonal base that has the highest atomic packing factor

High-speed tool steel   High-carbon, high-alloy steels that have a strong matrix containing numerous carbides; used for machining

Homogenization   A high-temperature heat treatment to reduce inhomogeneous composition

Hot melt   An adhesive that is melted for adherence

Hot shortness   Crumbling of steel during forging caused by low melting iron-sulfur grain boundary film

Hot working   Deformation carried out above the recrystallization temperature

Hot work tool steel   Tool steel used for hot-working metals

HSLA steels   Acronym for high-strength low-alloy steels

Hume-Rothery rules   Rules for substitutional solid solubility of alloy elements

Induction furnace   A melting furnace that has a water-cooled copper solenoid for high-frequency induction melting

Impact   A method for measuring the toughness of a material whereby the specimen is struck with a kinetic force

Impurity   An unwanted element in small quantity

Ingot   The form of a cast metal that will be deformed

Inhibitors   Additives that prevent corrosion

Inspection   An examination for specific information, such as an examination for corrosive damage

Interstitial   An atom that fits into the hole, or interstice, of a crystal lattice

Invar   An fcc Fe-36% Ni alloy that has zero thermal expansion up to its Curie temperature

Invariant reaction   A phase change that can only take place with a thermal arrest

Inverse lever rule   The relation between amount and composition of phases based upon material balance

Ionic bond   A bond between ions of elements based upon electrostatic attraction

Iron–iron carbide phase diagram   The portion of the Fe-C phase diagram that is used for the study of steel

Isothermal   A reaction that takes place at constant temperature

Kovar   A controlled thermal expansion alloy of iron, nickel, and cobalt used for glass-to-metal seals

Lap joint   A metal joint design where two layers overlap each other

Laser welding   A welding process using laser energy

Lattice   The three-dimensional array of atom positions that make up crystals

Lead-tin phase diagram   A binary equilibrium phase diagram of alloys containing lead and tin

Liquidus   The line of an equilibrium phase diagram above which only liquid is present

Macroporosity   Porosity of a cast metal that has substantial size

Magnesium alloys   Alloys that have magnesium as the major element

Magnetic domains   Magnetic areas that have a single dipole moment

Magnetic particle test   A nondestructive test for welds whereby ferromagnetic particles are attracted to weld defects

Magnetic properties   The properties of a material that are unique to magnetization characteristics

Magnetic saturation   The maximum magnetic field induced in a material by a smaller applied field

Magnetostriction   The stress effect of magnetization

Malleable cast iron   A form of cast iron with graphite particles formed from cementite in white cast iron

Mandrel   A metal core around which other metal can be shaped

Manual process   Any process controlled by human manipulation

Maraging steel   A carbon-free iron alloy containing nickel and other elements that forms a martensite that can be aged, that is, hardened by precipitation

Martempering   A heat treatment to avoid quench cracking where austenite is quenched to above the martensite start temperature, then slowly cooled to form martensite

Martensite   A hard, single iron-carbon phase formed by diffusionless transformation of austenite upon quenching

Material design   A consideration for corrosion protection

Material selection   Choice of materials based upon design and properties needed for an application

Mechanical joining   Joining of materials using mechanical fasteners such as bolts or rivets

Mechanical properties   The properties based upon mechanical deformation of a material

Metallic bond   Atomic bond based upon electron sharing with distant as well as nearest neighbor atoms; a mobile covalent bond

Metalworking   Metal deformation processes such as forging, rolling, extrusion, and the like

Microporosity   Microscopic porosity of a cast metal

Microstructure   The distribution and size of phases of an alloy as seen in a microscope

Mild steel   A ductile low-carbon steel

Monel   A nickel-base alloy containing copper

Muntz metal   A copper-zinc alloy containing 40% zinc

Music wire   A plain carbon steel containing about 0.80% carbon

NDT   The acronym for nondestructive testing, usually of welds

Necking   The reduced area of a deformed metal sample prior to ductile fracture

Neutron   Subatomic paired proton and electron particle that is in the nucleus

Nichrome   An alloy of nickel and chromium that has high electrical resistance

Nitriding   A method of surface hardening steel that forms nitrides with alloying elements in the steel

Noble   Not corrosive; a cathode in an electrochemical cell

Nonferrous   An alloy that does not contain substantial amounts of iron

Nonmetallic inclusion   An impurity such as an oxide or sulfide that is entrapped in a solid metal or alloy

Normalizing   Austenitization of steel followed by air cooling

Nucleation   The first stage of a phase transformation, for example, nucleation of new grains in recrystallization

Oxyacetylene weld   A weld made by the energy from combustion of acetylene

Oxygen concentration cell   A galvanic corrosion cell in which the oxygen-depleted electrode is anodic

Paramagnetic   Relating to the small or nonexistent magnetization of a material in an applied magnetic field

Passivation   Method for improving corrosion resistance of a metal by treatment of the metal or the environment

Passivity   Behavior of a metal that is more resistant to corrosion than expected

Pearlite   Two-phase lamellar eutectoid structure of iron and iron carbide

Peritectic   An invariant reaction in which solid and liquid phases form a single solid of different composition upon cooling

Peritectoid   An invariant reaction in which two solid phases form a single solid of different composition upon cooling

Penultimate grain size   The grain size following hot work or preceding cold work

Periodic table   The orderly listing of the chemical elements

Permanent magnet   A ferromagnetic material that has been magnetized and remains magnetized at remanence; also referred to as *hard magnet*

Permeability   A nonlinear property that relates the induced magnetization to the applied magnetic field

pH   A chemical measure of acidity

Phase   Chemically homogeneous portion of a microstructure

Phase diagram   The temperature-composition diagram that describes the relationships of phases at equilibrium

Physical properties   Properties of a material that do not reflect a mechanical or chemical change

Pitch   The distance between rivets in mechanical joining

Plain carbon steel   Steel with no major alloy additions other than carbon

Plasma welding   A welding process using plasma energy

Plastic deformation   Change in shape of a solid metal caused by an external force

Polarization   Change in charge of an electrode brought about by a current passing between the anode and cathode

Polynary alloys   Metal alloys containing more than three major elements

Precipitation hardenable stainless steel   A stainless steel that can be strengthened by a precipitation heat treatment

Precipitation hardening   Strengthening by heating a supersaturated solid solution, causing an increase in strength due to precipitation of a new phase

Preforms   Materials such as solder that have been preshaped to promote productivity and conserve materials

Press brake   A versatile machine for plastic deformation that compresses or bends metal between dies

Primary process   A process that converts an ingot into a billet prior to final shaping

Process anneal   A recrystallization anneal to soften for the purpose of further deformation processing

Protons   Subatomic positive particles in the nucleus

Projection weld   A resistance weld where one adherend is raised (projected) to increase resistance

Punch and die   The parts of a mold for compressing material

Quench   Rapid cooling by inserting a hot part into a cold medium such as water or oil

Radiography   A nondestructive analysis of weld joints using X rays

Recrystallization   The complete replacement of cold-worked metal by a new set of grains caused by heating to the temperature known as the recrystallization temperature

Reduction in area   The change in cross-sectional area divided by the original area of a mechanical test specimen, usually measured to fracture

Remanence   The magnetization remaining in a ferromagnetic material when an applied magnetic field is removed

Resistance   The measured electrical property of a metal of specific dimensions, equal to voltage divided by current

Resistivity   The property of a metal that resists flow of electrical current

Resolved shear   The actual shear force caused when a force is applied at an angle

Rolling   A metalworking process that compresses metals by passing them between rolls

Resistance welding   A welding process in which resistance heating of the metal provides the energy

Riser   A feedhead in a foundry mold to compensate for solidification shrinkage

Rivet   A mechanical fastener

Sacrificial anode   An electrode that preferentially corrodes, thus protecting an important part

Screw dislocation   The linear defect caused by shearing part of the crystalline lattice

Seam weld   A resistance weld formed by moving wheel electrodes

Secondary process   A metalworking process that forms a part near its final shape

Secondary recrystallization   A high-temperature phenomenon that occurs in thin fcc sheet metals that have been heavily deformed then fully recrystallized; a second new set of macroscopic grains is formed

Segregation   Chemical inhomogeneity; usually refers to solidification microstructures

Sensitization   Precipitation of carbides during heat treatment of austenitic stainless steel that reduces the amount of chromium below that needed for corrosion resistance

Shearing   Cutting actions of two blades moving opposite each other in close proximity

Shear strain   Elastic displacement produced by shear loads

Shear stress   Load per unit area, using the area parallel to the load

Shock-resistant tool steel   A tool steel with high toughness

Shrinkage   The volume change upon cooling; usually associated with solidification

Slip   Movement of atomic planes over one another

Slip system   The combination of slip plane and slip direction that enables plastic deformation

SMAW   The acronym for shielded-metal arc welding

Soft magnet   A ferromagnetic material that has high permeability, low coercive force, and low remanence

Soldering   A metal-joining method that uses a dissimilar metal alloy that bonds to the adherend to form a microalloy

Solid solution   The mixing of a solid solute metal in a solid solvent metal to form a single-phase alloy

Solidification   The phase change from liquid to solid

Solidus   The line of an equilibrium phase diagram below which only solids are present

Solutionizing   Heating an alloy into a single phase region for homogenization as the first step in precipitation-strengthening heat treatments

Solvus   The line of a binary phase diagram that indicates the limit of solubility for a single phase

Spheroidization   A heat treatment of white cast iron to form the graphite spheroids of malleable cast iron

Stabilizer   An alloy additive that stabilizes a phase with respect to temperature or composition

Stainless steel   An iron-chromium alloy that contains sufficient chromium to have superior resistance to corrosion

Stamping   A forming process that uses a punch and die for bending and/or shearing a material to final shape

Steel   The alloy of iron and carbon usually containing less than 1% carbon

Steelmaking   The process of making steel from iron ore

Stickscrew   A formed length of screws designed so that the screws separate at a specified torsion; used for semiautomated assembly

Strain   The change in length or cross section with respect to original dimensions, caused by a load

Strain hardening   The strengthening of a metal or alloy caused by cold work

Strain ratio   The ratio of a metal's ability to be deformed in the plane of a sheet to its ability to resist deformation in the thickness of the sheet

Stress   The load per unit area applied to a shape

Stress cell   A galvanic cell in which the stressed area becomes anodic to unstressed areas

Stress concentration   The intensification of any applied stress relatable to geometric factors, usually related to crack propagation and failure

Stress relief anneal   A low-temperature anneal to remove residual stress in a metal

Substitutional solid solution   A solution in which solute atoms replace solvent atoms in the crystal lattice

Superalloys   Alloys with high strength at high temperatures

Surface hardening   Any method that increases the hardness of the surface of a metal with respect to its interior

Temper   To reheat quenched steel to convert martensite to ferrite and cementite in a microstructure called tempered martensite

Temper carbon   Irregular shaped graphite formed in malleable cast iron

Tension   The stress on a material caused by opposing forces that stretch the material

Texture   Anisotropy of deformed metals

Thermal arrest   The length of time for an invariant reaction to proceed

Thermit weld   A weld formed by heat from the chemical reaction of the filler metal

Thermocompression weld   A weld formed by the combination of temperature and pressure

Thermocouples   Devices for measuring temperature

Thermostats   Devices for activating equipment at specific temperatures

Tie lines   Constant temperature lines between solvus, solidus, and/or liquidus lines in two-phase regions of phase diagrams

Titanium alloys   Alloys of titanium, usually with aluminum, vanadium, and/or tin

Toughness   The combination of strength and ductility, usually measured by impact testing

True strain   The change in length or cross section with respect to actual length or cross section

True stress   The change in load with respect to actual cross-sectional area

TTT diagram   The diagram describing isothermal transformation of austenite at temperatures below the eutectoid

Turk's head   A changeable set of rolls that can produce square or rectangular cross sections when metal is pulled through them

Ultimate tensile strength   The highest engineering stress that a metal can withstand without fracturing

Ultrasonic test   A nondestructive test for welds that utilizes the reflection of ultrasonic waves to detect defects

Unit cell   The smallest entity that can be repeated to generate a crystal structure

Upset forging   The compression of a solid between two open (flat) dies

Vacancy   A defect where an atom is missing from its proper lattice site

Vacuum melting   Melting of metals in a vacuum to prevent contamination

Valence   The electronic charge of an ion

Visual inspection   A nondestructive, careful examination of metal joints

Wave soldering   An automated soldering process whereby the parts to be joined are passed through a wave of molten solder

Welding   A metal-joining process whereby the adherends are melted and become part of the joint

Yellow brass   See Cartridge brass

Yield strength   The highest strength that deforms a metal only elastically, sometimes referred to as the flow strength because higher values will plastically deform the metal

Zinc alloys   Alloys that have zinc as the major component

Index